D0122177

# The Keys of Egypt

# The Keys of Egypt

## THE OBSESSION TO DECIPHER EGYPTIAN HIEROGLYPHS

### Lesley and Roy Adkins

HarperCollins*Publishers*

THE KEYS OF EGYPT.    Copyright © 2000 by Lesley and Roy Adkins
All rights reserved. Printed in the United States of America. No part
of this book may be used or reproduced in any manner whatsoever
without written permission except in the case of brief quotations
embodied in critical articles and reviews.
For information, address HarperCollins Publishers Inc.,
10 East 53rd Street, New York, NY 10022.

HarperCollins books may be purchased for educational, business,
or sales promotional use. For information, please write:
Special Markets Department, HarperCollins Publishers Inc.,
10 East 53rd Street, New York, NY 10022.

FIRST EDITION

*Set in Photina by Nigel Strudwick*
*Hieroglyphic font by Cleo Huggins*

Printed on acid-free paper

Library of Congress Cataloging-in-Publication Data
has been applied for.

ISBN 0-06-019439-1

00 01 02 03 04 RRD 10 9 8 7 6 5 4 3 2 1

*To Liz, Jon and Poppy, with love*

# Acknowledgements

It is a pleasure to acknowledge the help of many people and organizations during the writing of this book. We would first of all like to express our sincere thanks to L'Asiathèque for permission to quote from the letter to Angelica Palli published in *Jean-François Champollion: Lettres à Zelmire* by Edda Bresciani (1978). Part of the 'Teaching of the Vizier Ptahhotep' and the words 'It is good to speak to the Future; it shall listen,' both originally published in *The Tale of Sinuhe and Other Ancient Egyptian Poems 1940–1640 B.C.* translated with an introduction and notes by R. B. Parkinson (1997), are reproduced by kind permission of Oxford University Press. We are also grateful for permission to reproduce part of a letter written by Heqanakht, the spell to protect a child, and the eulogy to dead authors, all originally published in *Voices from Ancient Egypt: An Anthology of Middle Kingdom Writings* by R. B. Parkinson (1991), which are © the British Museum, British Museum Press. The spell is reproduced after the version in *Voices from Ancient Egypt*, kindly amended by R. B. Parkinson.

All the illustrations are the copyright of Lesley and Roy Adkins Picture Library, except for the photograph of the Rosetta Stone which is reproduced by the kind permission of The British Museum, and the photographs of the young Jean-François Champollion and Jacques-Joseph Champollion-Figeac, which are taken from *Les Deux*

*Champollion* by Aimé-Louis Champollion-Figeac (1887) and reproduced by the kind permission of The British Library.

The staff of many libraries gave us invaluable assistance, most
notably the London Library; the University of Bristol Arts and
Social Sciences Library and Worsley Chemical Library; the Griffith
Institute Library and Bodleian Library in Oxford; the British
Library; Bernard Nurse and Adrian James of the Society of Anti-
quaries of London's Library; David Bromwich of Somerset Studies
Library; and Marie-Françoise Bois-Delatte at the Bibliothèque
municipale in Grenoble. Also in Grenoble we are pleased to
acknowledge the assistance of Jean-William Dereymez and the
Société Champollion, and M. and Mme. Chateauminois in nearby
Vif were very obliging. At Figeac our stay was made particularly
enjoyable and useful by the assistance of Madame Prévôt at the
Musée Champollion, the staff of the municipal library, the Château
du Viguier du Roy and Taxi Fricou.

Special thanks must be extended to Dr. Richard Parkinson of the
Department of Egyptian Antiquities at the British Museum, who
has been most generous in sharing information. We are indebted to
Dr. Nigel Strudwick for his very useful help and meticulous typeset-
ting of the text and hieroglyphs. We must also not forget Gill and
Alfred Sims for their practical help.

Within HarperCollins thanks are due to Larry Ashmead,
Michael Fishwick, Kate Morris, Sonia Dobie and Chris Bernstein for
all their invaluable help.

Our overwhelming gratitude is kept finally for Patrick Walsh,
𓋹𓏤𓃮𓈖𓏭𓏛, without whom this book would most certainly
never have been written.

# Contents

# List of Illustrations

# The Keys of Egypt

# (The Beginning of Time)

The house at 28 rue Mazarine, where Jean-François Champollion lived and carried on his research into hieroglyphs, was less than 200 yards from the Institute of France where his brother Jacques-Joseph had his office. Towards midday on 14 September 1822, Champollion covered the distance in the shortest time possible. Clutching his papers, notes and drawings, he fled along the narrow, gloomy street, around the corner and into the Institute. Not fully recovered from his latest spell of ill-health and at the highest pitch of excitement, he was already breathless as he burst into his brother's office, flung his papers on to a desk and shouted, '*Je tiens l'affaire!*' ('I've found it!'). Working since early morning on the latest drawings of inscriptions from Abu Simbel, he had at last seen the system underlying the seemingly unintelligible Egyptian hieroglyphs, and it was now only a matter of time before he would be able to read any hieroglyphic text. He began to explain to Jacques-Joseph what he had discovered, but only managed a few words before collapsing unconscious on the floor. For a few moments his brother feared he was dead.

Perhaps not quite in the way he had always hoped for, this was to prove the most important turning point in Champollion's turbulent life. Through years of ever-increasing preoccupation with hieroglyphs he had been working towards this goal, but his first tentative steps had been made before he had chosen his life's work and even before he had seen any hieroglyphs – he was drawn to his destiny by an insatiable curiosity about the origins of the world. Early in childhood, apparently neglected by his parents, he was looked after and to some extent spoiled by his brother and three sisters, and they doted on the bright baby of the family who was so much younger than themselves. Champollion's high intelligence and extraordinary genius for languages were recognized by his brother, who was determined that these gifts should not be wasted. Having had his own schooling curtailed by the terrible upheaval of the French Revolution, Jacques-Joseph resolved to minimize its effects on Champollion, initially by giving him lessons, since all the schools had been shut down. Later a private tutor was found for the boy, but as a precarious political stability returned under the ascendance of Napoleon, the schools reopened. By the age of twelve Champollion was so prodigiously proficient in Latin and Greek that he was allowed to begin studying Hebrew, Arabic, Syriac and Chaldean. Since his knowledge of Latin and Greek had already opened to him a world of books on all manner of subjects, his passion for Oriental languages initially appears a curious caprice on the part of this son of a rural bookseller, born in the remote town of Figeac in southwest France. In reality Champollion had already decided to take on one of the great intellectual challenges: to investigate the creation of the world and the beginning of time itself.

Although the Revolution had outlawed the Catholic Church and suppressed religion, the only model for the origin of the world was still contained in the Old Testament of the Bible, which was believed to be a description of the history of the earth from the time of its creation by God. Scholars preparing to examine this theory had to possess a good knowledge of Oriental languages in order to study early versions of the Biblical texts and related documents. Since it was still believed that people lived on the earth very soon after the world began, it was natural to use the tools of history and philology to look for its origin – archaeology and geology were only in their infancy, not yet respectable sciences. Champollion's insatiable curiosity was to tempt him towards various other fields of study on many occasions, but once he became aware of the potential of ancient Egypt, he found the focus for which he was searching. He was gripped by energetic enthusiasm for this mysterious country, a Biblical land whose history was interwoven with that of the Israelites, but the history of Egypt (indeed, virtually all knowledge of Egypt) was locked away in hieroglyphic texts that could not be read – texts that might contain unimaginable secrets, even an accurate account of the origin of the world. Here was a challenge worthy of his talents, a prize of untold knowledge, forgotten for centuries – if he could only decipher the hieroglyphs.

Apart from his exceptional gift for languages, another gift that was to prove decisive in Champollion's success was his extraordinary visual memory, which allowed him to pick out similar signs and groups of signs among the thousands of hieroglyphs he was to study. It may have been this visual memory that caused his initial problems with writing and spelling – as a child he seems to have seen words as pictures and pictures as words, making little distinction between writing and drawing. This unconventional and careless approach was probably a result of his early childhood when he tried to teach himself by copying words from books – an indication of his ability to tackle problems in his own original way. In the unfocused freedom of these formative years, with no proper teaching, he developed the wide-ranging curiosity that later provided both the main driving force of his life and a tendency to be distracted by ever more interesting irrelevancies, but the legacy of this unusual childhood was not altogether beneficial. Restricted to the home because the social unrest of the Revolution made the streets unsafe for children, Champollion at least had the freedom indoors to explore whatever caught his attention, but this later caused problems when he was forced to cope with the disciplines of the schoolroom and the necessity to study subjects, such as mathematics, that completely failed to interest him. It took him many years to learn how to cope with life as an ordinary schoolboy and he never fully adjusted to it, simply because he was far from ordinary. With a keen sense of humour, he increasingly used his flair for satire and biting wit to defend himself as he strived to survive the rigours of school. With friends and family, though, he was invariably kind and generous.

The violent mood swings and tantrums born of the frustrations of Champollion's early school-days gradually gave way to a grudging tolerance of lessons that were incomprehensible or numbingly repetitive and of teachers who were more provocative than inspiring, as he tried ever harder to suppress his outrage at injustice and disguise his boredom. What he did not hide was an ability and a passion for those subjects which did interest him. Initially an amusement, drawing was a skill that he continued to develop and would become essential in his study of hieroglyphs, and botany was an enthusiasm that never entirely left him, but to his aptitude and obsession for languages was added an increasing immersion in ancient history.

As his education progressed, it was the work done in his spare time that began to display the developing skills that would be so important for his study of hieroglyphs. Once committed to a project Champollion was capable of the most meticulous study, patiently and painstakingly examining and assessing all available sources of evidence. He had a passion for listing, classifying and analyzing the accumulated material before using strictly logical reasoning to produce his results. Above all, he was stubborn – he might be forced to leave off a project or stop it altogether, he might be sidetracked or hindered a hundred times, but he never gave up. He also had the courage and independence to approach problems with an open mind. Born and brought up in the time of the French Revolution, when religion had been officially abolished and actively suppressed, his education was nevertheless often in the hands of men who were devout Catholics (many of them priests and monks before the Revolution), enabling him to develop a flexibility of thought that was to prove crucial in understanding hieroglyphs. Where his rivals tended to be factional and polarized (for or against the Church, for or against Napoleon), and equally rigid in their intellectual theories, Champollion weighed the evidence and drew his own conclusions. This approach was both a blessing and a curse: applied to academic problems it was essential, but in a time of political upheaval it could prove fatal. Initially a critic of Napoleon, Champollion became a fervent supporter in the weeks before Napoleon's final abdication – an unfortunate decision that brought him under immediate suspicion from the restored monarchy and lifelong hatred and opposition from many Royalists.

The final element in Champollion's success was the availability of material. The number of hieroglyphs that previous scholars had at their disposal was severely limited, being mainly derived from Egyptian monuments and artifacts that had been imported into Europe long ago. Because Egypt had been closed to outsiders for centuries, attempts to decipher hieroglyphs had ground to a halt as insufficient source material made the task impossible. Yet even as Champollion was attending his first formal lessons, Napoleon was conducting a momentous military campaign in Egypt that would eventually bring all things Egyptian to the notice of western Europeans and especially the French. Napoleon's campaign in Egypt failed in its military objectives, but the savants accompanying the expedition took back to France a mass of notes, drawings and artifacts that were to amaze the scholars of Europe, and the soldiers who survived the campaign returned with stories of an exotic land of stark contrasts – stories that would have been enlarged and embellished each time they were told. From Napoleon himself down to the lowest ranks of conscripts, everyone who had taken part in the Egypt expedition was deeply affected by the experience, and a fascination for Egypt (in effect a newly discovered country) spread throughout France, creating a wave of Egyptomania. Over the following decades, the Egyptomania faded but the French affinity for Egypt, the colony France never had, has continued to the present day.

By the time Champollion arrived in Paris in 1807, the city was home to the most brilliant linguistic scholars in Europe. It also possessed a mass of exciting but as yet barely studied material just brought back from Egypt by Napoleon's expedition, and libraries were choked with precious books and manuscripts that had been looted from all over Europe by Napoleon's armies. The attempt to decipher hieroglyphs which Champollion had already begun soon developed into a race between himself and his rivals – a strange race run in darkness, where each competitor was often unknown to the others until he chose to reveal publicly what strides he had made. Scholars from all over the Continent began to study hieroglyphs, with many concentrating on the inscriptions of the newly discovered Rosetta Stone, whose three texts held out the hope that the hieroglyphic text could be matched with the Greek text to provide a key to translation. As more and more people joined in, the serious competition resolved itself into a duel between two men, and in a curious reflection of the politics of the time, one was French and one was English – Jean-François Champollion and Thomas Young. The competition was open and anyone could try his hand at decipherment, but there was no formal race, no prize was offered, no money, no medals – above all, there were no rules. Yet each would-be decipherer fully understood what he was striving for: a place in history, a reputation as the one who liberated ancient Egypt from the ignorance and obscurity that had come to surround it – the immortal acclaim due to the man who first deciphered hieroglyphs.

# CHAPTER ONE

# (The Land of Egypt)

Josephine never did see Egypt. She begged Napoleon to take her with him, but for once he was undecided. He knew that his expedition to Egypt was a gamble – if the French fleet, laden with troops, supplies and armaments, was caught by the British Navy, there would be little chance to fight or flee. Had Josephine accompanied Napoleon, she would have been one of the first western women to see Egypt in more than a thousand years, because conditions were so dangerous that travel up the Nile Valley was only for the fearless, the foolish or the suicidal.

On 19 May 1798 General Napoleon gave orders for the French fleet to set sail, finally deciding not to take his wife Josephine, but to send for her once his expedition had successfully evaded the British. He had, in fact, given orders that no women, other than the few officially authorized ones such as laundresses and seamstresses, were to travel with the expedition, but it was not unusual for women to accompany their husbands and lovers on military campaigns, and in the event Napoleon's orders were not strictly observed. Some wives of officers travelled with them openly, while other women stowed away on the ships or were disguised as men. In all, about 300 women sailed to Egypt.

The expedition had only been at sea for four days when Napoleon decided to take the risk and send for Josephine after all. The frigate *Pomone* was despatched to collect her, but by the time it reached its destination, Josephine was too ill to travel. Having stayed to see Napoleon sail from Toulon, she had made her way to the spa town of Plombières in Lorraine to take the waters, but on 20 June she suffered a serious accident when a wooden balcony collapsed fifteen feet onto the street below. For three months Josephine endured a long and painful recovery, while the local physician treated her with boiled potatoes, brandy and camphor applied as compresses, along with leeches, hot baths and frequent enemas. When she was again able to write, she could only lament in letters to her friends that she was unable to sail for Egypt: 'I have received a charming letter from Bonaparte. He tells me that he cannot live without me, to come and rejoin him, and to go to Naples to embark. I very much wish that my health might allow me to leave immediately, but I see no end to my cure. I cannot remain standing or sitting for ten minutes without terrible pains in my kidneys and lower back. All I do is cry.'

This accident was a critical moment for Napoleon and Josephine – by the time she was well enough to travel, Napoleon had been given proof of her adultery and no longer wished to see her in Egypt. At one time bewitched by this woman, six years older than himself, whom he had married two years earlier, his relationship with her was never the same again, and shortly afterwards Napoleon took the first of a series of mistresses: the newly married wife of an army lieutenant, Pauline Fourès, who had accompanied her husband disguised as a soldier and who came to be known among the troops as Cleopatra – the name that, written in both Greek letters and hieroglyphs, would later provide one of the vital keys to the decipherment of hieroglyphic writing.

Napoleon Bonaparte, born in 1769 at Ajaccio in Corsica of minor nobility, had received a military education in France and entered an artillery regiment of the French army in 1785. The French Revolution, which broke out four years later, would lead to war with many European states fearful of the spread of social reform, and from 1796 Napoleon led the army to stunning victories over the Austrian forces in Italy. For a brief period France was only at war with its implacable enemy, Britain, but Napoleon judged any further attempts to invade that country as too perilous without control of the sea. Instead, he looked to destroy Britain by seizing Egypt: French control would disrupt British trade with its richest possession, India, and provide a base for military expeditions to the subcontinent.

It also suited Napoleon to be far away from politics in Paris at this time. His hope was to return from Egypt in triumph and take control of a coup d'état that others were already plotting: for their part, the members of the Directory (a committee of five directors holding executive power under the Revolutionary constitution of 22 August 1795) were glad to see him go, expecting the expedition to fail and put an end to the ambitious young general's political career – when Napoleon suggested that he should mount the expedition they rapidly consented.

By the time his expedition reached Egypt nearly six weeks after leaving France, the country had been part of the empire of the Ottoman Turks for nearly three centuries – the Turks had taken control from the Arabs who had themselves invaded Egypt nearly 900 years earlier. Before Napoleon, only a handful of travellers – invariably men – had ever ventured south of the Nile Delta. The small number of merchants were largely confined to Cairo, Alexandria, Rosetta and Damietta, and the main centre for those from the west was Cairo, where they had their own walled quarter, the entrance to which was guarded by Turkish soldiers. Even in the Nile Delta, it was not safe for westerners to travel outside these settlements without an armed escort, while travel south of the Delta was not even considered. As a consequence, the fifty to sixty French merchants living in Egypt were unable to provide much information about the country: when Napoleon and his generals arrived they realized just how little they knew about the land they had come to conquer.

Through sheer good fortune the ships carrying the expedition reached the coast of Egypt at the end of June 1798, having eluded a powerful British fleet that was combing the Mediterranean to locate and destroy them. The expedition had a military force of some 38,000 troops on board 400 transports, with sixty field guns and forty siege guns, but only 1,200 horses for around 3,000 cavalrymen because Napoleon expected to use camels for transport. There was also a party of savants: despite having been invited on a tropical voyage without knowing the destination, over 150 members of the National Institute had been willing to join the expedition. This Institute, established in Paris in 1795, included eminent men in all branches of science, and Napoleon was very proud that he had been elected to the Institute in 1797, a fact which probably helped persuade so many of these savants (officially listed as 167 in number) to take such an extraordinary leap in the dark. If the British had found and destroyed the French fleet, the cream of France's intellectual and artistic talent would have been lost: for this reason, the savants travelled on at least seventeen ships, with each specialist group split amongst different ships.

The specialist groups of savants included astronomers, civil engineers, draughtsmen, linguists, Orientalists, painters, poets and musicians, with celebrities such as the brilliant mathematician Jean-Baptiste-Joseph Fourier, the scientist and mathematician Gaspard Monge who had invented descriptive geometry, and the chemist Claude-Louis Berthollet. Amongst the other notable scholars were the inventor and balloonist Nicolas Conté, who is probably best known for inventing the graphite pencil, Déodat Gratet de Dolomieu the mineralogist, after whom the Dolomite mountains were named, the naturalist Geoffroy Saint-Hilaire, the artist and engraver Dominique Vivant Denon, the poet François Auguste Parseval Grandmaison, and the engineer and geographer Edme-François Jomard.

Napoleon's true motives in taking a large group of talented civilians on such a perilous military adventure are not known, but their presence allowed him to claim that the expedition was a civilizing mission, not one of imperial conquest. There were even plans to cut a canal through the Isthmus of Suez to join the Mediterranean to the Red Sea and so provide a new sea route to the East. This aspect appealed to Napoleon, who saw himself as following in the footsteps of Alexander the Great, a Macedonian Greek who took control of Egypt in 331 B.C. before setting out on his campaign of conquest through the Persian Empire to India and Afghanistan, beyond the Hindu Kush. Having died of poison or fever at Babylon, Alexander was brought back for burial in his newly founded city of Alexandria in Egypt. He had taken a group of scholars and scientists on his Persian campaign, and for centuries afterwards all European scientific knowledge of the East was based on the information that these men had gathered: Napoleon's group of savants may well have been designed to equal or surpass Alexander's.

It is with the few savants who became interested in the monuments of ancient Egypt that the story of deciphering the hieroglyphs really begins. The entire group of savants was expected to record every aspect of the country, including geology, hydraulics, fauna and flora, religion, agriculture and manufacturing – a task completely superfluous to the stated aims of the military expedition. The recording of ancient monuments was not envisaged as a major concern, because the savants were largely unaware of the huge number of ancient monuments that survived. If there was any immediate practical benefit to be gained from the presence of the scientists and engineers it was to assess and record the country's current wealth, strategic value and possibilities for development, with a view to making Egypt a colony of France – a practical impossibility given the political disruption in France at that time. In reality, the presence of the savants in Egypt was hardly more than a whim on the part of Napoleon who perhaps dreamed of transcending the feats of Alexander, yet without that whim Egyptian hieroglyphs might still be undeciphered. It was the return to France of those savants who survived their stay in Egypt, along with thousands of drawings of hieroglyphs on the walls of tombs and temples,

that not only sparked renewed interest in deciphering the hieroglyphs but provided for the first time a mass of material for their
study – material that took three perilous and gruelling years in
Egypt to gather. It was also the start of the French passion for Egypt.

Having safely reached Egypt, no time could be wasted in getting
ashore, because the French knew that British warships might
appear at any time to attack before the army was disembarked and
before the French fleet could be reorganized into battle formation.
The landing began off Marabout beach, to the west of Alexandria,
at about midday on 1 July, but the ships were some three miles off-
shore, with rocks and reefs between them and the beach, and the
weather was rapidly deteriorating. The first boats reached land at
eight in the evening, and the troops continued disembarking all
through the night. Because of the heavy surf and the need to avoid
obstacles, it took up to eight hours to row the boats from the ships
to the shore, and many men were injured or fell overboard while
transferring from the ships to the boats in the rough seas. Napoleon
recorded only nineteen men drowned during this operation, but it
is likely this figure is more propaganda than accurate record and
that casualties were far higher.

Disembarkation of the army was not completed until 3 July, but
Napoleon did not wait for this. At dawn on 2 July, walking at the
head of a column of around 5,000 troops, he began the march on
Alexandria, despite the fact that no artillery, horses or even drinking water had been landed from the ships. The tired and hungry
soldiers carried nothing but their weapons and the clothes they
marched in. No road existed from the landing place to Alexandria,
and the few wells and water cisterns had been sabotaged by
nomadic Bedouin Arabs who also continually harassed the French
and captured any stragglers. Once they became known, the atroci-
ties committed on prisoners by the Bedouin discouraged soldiers
from straggling on subsequent marches. The French reached the
outskirts of Alexandria at eight in the morning, and despite the
troops being exhausted and suffering from extreme heat and thirst,

Napoleon ordered an immediate attack. The inhabitants were poorly armed, terrified of the approaching army, and had spent the night sending messages to Cairo begging for reinforcements. The French, desperate for water and facing only feeble resistance, had control of the city in under three hours.

It was not until 4 July that the savants disembarked. They had been regarded as a low priority, and some had not been treated very well after the main force had left the ships, being made to sleep on deck and denied food. Most were unceremoniously dumped on the outskirts of Alexandria, along with their personal luggage, and were left to fend for themselves. Alexandria was a shock and a disappointment. There was no trace of the city that had once been the cultural and intellectual centre of the ancient world, with its famous library of over 700,000 volumes, its temples, theatres, palaces and the tomb of Alexander the Great. The city had once measured one mile by three miles, and its twin harbours were protected by the famous Pharos lighthouse, built in the third century B.C. and regarded as one of the seven wonders of the world. Its population was reputedly over 300,000, possibly even one million, but from the time of the Arab conquest in the mid-seventh century Alexandria went into a steady decline, and due to earthquakes and subsidence much of the city became submerged. It is only now that the former glory of Alexandria is being brought to light by the work of underwater archaeologists. To the dismay of the savants, all they saw was a collection of rickety hovels, clustered around squalid narrow streets, inhabited by a population of less than 6,000.

The military hierarchy gave no thought to the savants, who spent several days finding shelter with Europeans already living in Alexandria and at the house of the British Consul who had left before the French arrived. It took a direct protest to Napoleon by the mineralogist Dolomieu before arrangements were made for provisions for the savants, and even then they only received ordinary soldiers' rations. Napoleon was intent on moving from Alexandria as soon as possible and so there was frenzied activity everywhere.

General Caffarelli, who was in charge of the savants, only had time for the military engineers among them and ignored the rest. As the campaign progressed, the savants became used to being treated like part of the army, but initially they resented being regarded as less important than the humblest of the troops and felt that being used as clerks and messengers (jobs Caffarelli assigned to the most vociferous complainers) was a waste of their talent.

Friction between the soldiers and the savants had already begun to develop during the voyage from France, with both sides complaining to Napoleon. Having a foot in both camps, he could see no reason for the discord and became impatient about the complaints. Indeed, he had made the situation worse by holding daily discussions, often on the deck of the ship, which the savants and his army officers were obliged to attend in order to discuss a wide range of subjects, supposedly in preparation for the Institute of Egypt that Napoleon intended to set up. The soldiers had various derogatory names for the savants, of which the most common one was 'donkeys.' On the march, the real donkeys that carried the baggage were jokingly referred to as 'demi-savants' (half-scientists), and when the ranks were formed into a defensive square before a battle, the bellowed order 'Donkeys and savants to the middle of the square' inevitably provoked laughter from the troops.

Temporarily established in Alexandria, the savants found nothing of interest in the squalid city and very few ancient remains to explore. The most obvious monument was Pompey's Pillar, a Roman stone column set on top of a hill, which dominated the town. Despite being named after the first-century B.C. general Pompey, whose severed head was presented to Julius Caesar when he landed in Egypt in pursuit of him in 48 B.C., the column was actually erected in the reign of the Emperor Diocletian (A.D. 284–305) to commemorate his visit to Alexandria – he was the last reigning Roman emperor ever to set foot in Egypt. Of more interest were Cleopatra's Needles, which were actually two obelisks, one still standing and the other fallen and half-buried in sand, both covered in hieroglyphs. These were the first real Egyptian monuments that the French encountered, but they had no connection with Cleopatra and were originally erected around 1500 B.C. in front of a temple in the ancient city of Heliopolis (now beneath the suburbs of Cairo). They were only moved to Alexandria in 10 B.C. by the Roman Emperor Augustus. The base of the standing obelisk had Greek and Latin inscriptions carved on it, but as it was buried in the sand these were not visible. Being unable to read the hieroglyphs, the savants were unaware that these monuments had already travelled the length of Egypt, from a quarry near Aswan far to the south, to be erected in Heliopolis and then re-erected in Alexandria. Decades after Napoleon the fallen obelisk was moved to the Thames Embankment in London, where it is still known as Cleopatra's Needle, and the standing obelisk was transported to New York and erected in Central Park.

The hieroglyphic inscriptions that the savants saw on the obelisks contained a high proportion of names within cartouches, such as,

which means 'Son of the sun god Ra, Ramesses, Beloved of the god Amun,' but they did not realize that such dedicatory inscriptions, consisting almost entirely of the names and titles of pharaohs, were not especially common, because this was the only type of hieroglyphic inscription already familiar to them, carved on the obelisks and other monumental sculptures in Rome which had been looted from Egypt by the Romans over 1,500 years earlier. It would be some months before the savants were fully aware of the richness of the ancient Egyptian remains that awaited them south of the Nile Delta, having so far only seen the desert and the disappointing remnants of Alexandria.

In stark contrast to the desert, the annual flooding of the river left a thick carpet of moist and fertile black silt along the Nile Valley that gave Egypt one of its ancient names:  (Kemet – the black land) – land so fertile that it produced the real gold of Egypt: 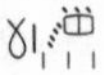 (grain). For thousands of years the annual flood cycle maintained a way of life which changed so very slowly from generation to generation that this change was imperceptible. The grain and other crops made the country so wealthy that it could sustain the huge labour force (conscripts rather than slaves) that was needed to build massive tomb complexes for their kings and vast temples for their gods, and as long as the gods regulated the annual Nile flood and made the sun shine each day, there was no incentive for change. Such an idyllic place did not escape the notice of its neighbours, and Egypt was often at war, eventually being invaded by a succession of enemies – Alexander the Great took possession of Egypt from the Persians, and the resulting Ptolemaic Greek ruling dynasty that followed him was in turn ousted by the Romans in 30 B.C. when Cleopatra VII was defeated and committed suicide. The Roman general Octavian conquered Egypt, and he returned to Italy to outmanoeuvre his enemies and become its first Roman emperor, at the same time making Egypt a province of the Empire.

For the Romans, the province was a fabulous prize – a land of strange gods and great wealth. The vast quantities of grain from the fertile Nile Valley were so important to Rome that Egypt was directly controlled by the emperor, and gold was so plentiful in Egypt that by comparison silver was an expensive import. Several subsequent Roman emperors visited the country and were fascinated by its ancient monuments: so much so that they transported back to Rome obelisks, sphinxes and various statues, all adorned with mysterious hieroglyphs. Egyptomania spread through Roman Italy, with tombs constructed in the shape of pyramids, and houses and gardens decorated in an Egyptian style. Some plain obelisks were even carved with fake hieroglyphs to make them appear more Egyptian, which would later confuse attempts at decipherment when Egyptomania spread through France as a direct result of Napoleon's expedition – once more Egyptian styles would be the height of fashion, and tombs marked by pyramids and obelisks would be built in Paris cemeteries just as they had been built outside the walls of Rome eighteen centuries before.

Under Roman rule the use of hieroglyphs had gradually declined and the rise of Christianity forced the abandonment of the pagan temples and the hieroglyphic script that was associated with them. The last dated hieroglyphic inscription was carved on a temple gateway on the island of Philae near Aswan in Upper Egypt on 24 August A.D. 394. After this fewer and fewer people could read hieroglyphs, and even simple inscriptions such as ⌒|○⊜⌒△ 𝕏⌒𝄇○ would not be understood until Champollion finally managed to decipher them.

Five days after arriving in Alexandria in July 1798, the savants were divided into three groups. Gaspard Monge and Claude-Louis Berthollet accompanied Napoleon, who left Alexandria on 7 July, leading a force to capture Cairo. The next day a group went with General Menou, who was going to Rosetta by sea, and the remainder stayed in Alexandria with General Kléber. Anxious to reach Cairo as quickly as possible, Napoleon had already sent forces led by Generals Desaix and Reynier on a desert route to Damanhur, intending to meet up with them there. The soldiers were not properly equipped for such a march, having only been provided with rations of dry biscuits and no water flasks. Only a few managed to find containers for water beforehand, and the march degenerated into a relentless search for something to quench their thirst.

From the moment the troops left Alexandria they were yet again harassed by bands of Bedouin, who kept up their assault all the way to Cairo. It was the Egyptian summer, the hottest time of the year, and although they started their marches before dawn, as soon as the sun came up the soldiers suffered from the blistering heat, made worse by the thick material of their uniforms and the lack of water to lessen their thirst. Those wells and cisterns not sabotaged by the Bedouin were quickly emptied – the troops with Desaix drank most of the water, often leaving none for those following on behind with Reynier. When a cistern with water was found, the troops fought to get at it, as there was seldom enough for everyone. With little or none to spare to soften up the dry biscuit, the troops suffered from hunger as well as thirst. Sometimes they would see a lake or water hole in the distance, lush with vegetation, and would run towards it only to see it vanish: desperate for water, and not having seen a mirage before, the troops were deceived time and time again. Gaspard Monge later studied mirages and managed to explain what caused them, but on the march from Alexandria to Cairo they caused such despair that many soldiers went mad and shot themselves.

About 18,000 men set out from Alexandria to Damanhur, and a few hundred were either killed by the Bedouin, committed suicide or died of heat and thirst. From Napoleon's point of view, the number of casualties was small, but the forty-five-mile journey was unending anguish for the troops, leaving at least one man dead for every 200 yards travelled, and one of the officers recorded that they left a trail of corpses behind them. The morale of the soldiers, already low after the sickening sea voyage, the dangerous landing and the assault of Alexandria, was destroyed by this march. With some justification, troops and officers alike blamed Napoleon and his administration for lack of foresight and inadequate supplies, and Napoleon became all too well aware that a crushing and lucrative victory over the Mamelukes was essential for restoring morale.

Within the declining Ottoman Empire, Egypt was still subject to the rule of the sultans from Constantinople, but had actually come to be dominated by the Mamelukes. *Mameluke* is the Arabic word for 'bought man,' and although the Mamelukes were bought as child-slaves, often from the Caucasus, they were trained as warriors and automatically became free men on receiving a military command. The Mamelukes formed the real aristocracy of Egypt, living in luxury on taxes squeezed from the rest of the population. Bands of Mamelukes under their local chiefs called beys terrorized the country and occasionally made war on the Ottoman Turkish army; the two foremost beys ruling on behalf of the sultan at the time of Napoleon's invasion were Ibrahim Bey and Murad Bey. Knowing nothing but horsemanship, killing and extortion, Mamelukes were capable of great personal courage, but could also retreat with spectacular speed – as Napoleon was soon to find out.

By 9 July all the surviving troops had reached Damanhur, and Napoleon led the combined force to El Rahmaniya on the River Nile. Although at its lowest annual level, the Nile was still a substantial body of water and the soldiers became hysterical with joy. They threw themselves into the water and revelled in it for hours; some died from drinking too much, too soon, after suffering so long from thirst. From here, Napoleon moved up the Nile to investigate reports that a force of Mamelukes under Murad Bey was approaching the town of Shubra Khit, eight miles to the south. A heavily built, cruel and cunning leader, Murad Bey lived for war and never admitted defeat – despite never being victorious. Napoleon was accompanied by a flotilla of gunboats on the Nile that had been requisitioned as transports at Rosetta and were designed to counter the Mameluke gunboats reported to be with Murad Bey's army. They also functioned as transports for non-combatants including Gaspard Monge, Claude-Louis Berthollet and Pauline 'Cleopatra' Fourès.

The French reached Shubra Khit on 13 July and for the first time confronted a Mameluke army. Although the Mamelukes relied on cavalry, they also had a force of infantry consisting largely of Egyptian peasants armed only with clubs. The French infantry formed up in squares, with cannon at the corners of each square, and what little cavalry they had was placed inside the square for protection. Puzzled by this formation, the Mamelukes nevertheless expected their usual headlong charge to be effective. Mamelukes were seldom captured in battle: relying on the speed of their attack and retreat, they were either victorious, retreating rapidly, or dead. Heavily armed with scimitars, javelins, maces, battle axes, daggers and carbines, and often carrying several pairs of pistols, the

Mamelukes were also richly dressed in brightly coloured silks and muslins, and each carried his personal fortune in coins and jewels. Their method of fighting was to charge, first firing their carbines and then their pistols, which they tossed behind them to be recovered by their servants. Throwing their javelins, they finally attacked with a scimitar, and some of them even gripped the reins in their teeth and used two scimitars, one in each hand.

For several hours the Mameluke cavalry circled the French squares, looking for a weak point to attack, but they only charged when the two opposing flotillas of gunboats started to exchange cannon fire. Once within range, the barrage of cannon, musket and pistol fire from the French squares repulsed the Mamelukes before they could do any damage, and after about an hour they retreated to their original positions. Meanwhile the French gunboats had been getting the worst of the battle, and even the civilians had joined in the fighting. Napoleon ordered his troops to assist the gunboats, and soon after a cannon scored a direct hit on the flagship of the Mameluke flotilla, which was destroyed in a spectacular explosion. This drew a burst of hysterical laughter from the French, and the Mameluke cavalry turned and fled, followed by the rest of the Mameluke army.

Winning the battle at Shubra Khit raised the morale of the French troops for a time, but Murad Bey and his army had escaped. The gruelling march towards Cairo was resumed, and the slow attrition of the French forces through heatstroke, thirst and suicide continued. On 20 July, as the French drew near to Cairo, they learned that Murad Bey had concentrated his forces on both sides of the Nile at Embaba, just north of the city, and the next day, after a twelve-hour march, the French reached Embaba at two in the afternoon, in the worst heat of the day. The pyramids, some ten miles away, were visible in the distance, and so the battle became known as the 'Battle of the Pyramids.' In his memoirs Napoleon recorded that he addressed his troops and pointing to the pyramids said, 'Soldiers, forty centuries look down upon you.' It is doubtful

whether many of the troops would have understood or cared about
the significance of the pyramids, but in any case the soldiers were
deployed over such a large area that only the few nearest Napoleon
would have heard him; this was more likely a remark to his officers
rather than an attempt to inspire his troops with an appeal to their
supposed desire for a place in history.

Formed up in squares, the French moved out of range of the
Mamelukes' entrenched gun emplacements and provoked the
Mameluke cavalry to charge. The French held their fire until
the Mamelukes were less than fifty yards away – when it came, the
volley of fire stopped the charge in its tracks. For another hour the
Mamelukes continued to charge the squares in vain, and finally fled
back to their entrenched positions just as these were being attacked
by the forward troops under Generals Desaix and Reynier. Confusion among the Mamelukes turned into a rout: Murad Bey escaped
with some of his cavalry, while the majority of the Mameluke infantry fled across the Nile. Marred only by the escape of Murad Bey, it
was the overwhelming victory that Napoleon had hoped for, and
the troops spent the next week fishing dead Mamelukes out of the
Nile to loot their gold, jewels and other valuables. The victory
proved to be the turning point that Napoleon had needed to regain
control of his disillusioned and potentially mutinous army. The
next day, 22 July, the leaders of Cairo sought to negotiate terms of
surrender from Napoleon, and two days later he entered the city.

From landing the first troops at the beginning of July 1798, it
had taken the French nearly a month to secure Alexandria and
Cairo, during which time the British fleet under the command of
Nelson continued to search the eastern Mediterranean for the
French expedition. Horatio Nelson had entered the navy in 1770
and served many years in the West Indies. Blinded in the right eye
at Corsica in 1794, he had lost his right arm at Tenerife three years
later, yet he was still a formidable rear-admiral and a master of
naval strategy. On 1 August Nelson arrived at Alexandria and, on
learning that the French fleet was anchored a few miles to the east

and with the wind in his favour, he immediately sailed on to Abou-kir. The French commander Admiral Brueys had placed the French transports and smaller ships in the harbours at Alexandria, but fearing the shallows and contrary winds there, he had anchored the largest warships, seventeen in all, in a curved defensive line across Aboukir Bay. The French sailors were primarily concerned with finding supplies, and over one-quarter of them were on shore, with some foraging as far away as Alexandria and Rosetta. Despite an appearance of strength, the ships were only prepared for an attack from the seaward side – the guns on the landward side were not manned and some were obstructed by stores and baggage.

It was two in the afternoon when the leading ships of Nelson's fleet rounded the headland of Aboukir. During the long search of the Mediterranean Nelson and his captains had found plenty of time to discuss tactics, and, to the surprise of the French, the British fleet immediately prepared to attack. The French captains, at a conference aboard the flagship *L'Orient*, were forced to scramble back to their ships. By four o'clock all fourteen ships of the British fleet had cleared the headland, and the battle began just two hours later, with little daylight left. As soon as he realized the danger, Brueys signalled for the men on shore to return and then prepared for battle, but his line of ships was anchored more than one and a half miles from shore – and more than half a mile from the shallows that would have protected the landward side of the French fleet. The leading British ships took the highly dangerous gamble of sailing down the landward side of the enemy line of ships, so allowing the British fleet to work its way down both sides of the French line, with two British ships engaging each French vessel as they went: the unengaged French ships farther down the line could do little to help.

The battle continued through the evening and night, with each French ship being hit by at least two broadsides for every broadside it managed to fire. On board the *Vanguard*, Nelson was suffering badly from toothache, and before the start of the action he had told his officers: 'Before this time to-morrow, I shall have gained a Peer-age, or [a tomb in] Westminster Abbey.' At about eight-thirty in the evening, he was struck in the forehead by a piece of iron, and a flap of skin hung down from the wound over his left eye, effectively blinding him. As he collapsed into the arms of Captain Berry, he murmured, 'I am killed; remember me to my wife.' The surgeon on board convinced Nelson that the wound was not fatal, and after it was treated he regained some of his sight and was able to go back on deck for a short while, where he witnessed the destruction of the French flagship L'Orient, which blew up with a flash visible in Alex-andria and a blast that was felt up to twenty-five miles away. Captain Berry wrote that: 'L'Orient blew up with a most tremen-dous explosion. An awful pause and death-like silence for about three minutes ensued, when the wreck of the masts, yards, &c. which had been carried to a vast height, fell down into the water, and on board the surrounding Ships.'

The battle was a complete victory for the British, with only two French warships escaping. French casualties were considerable – 1,700 killed and 1,500 wounded, and the French also lost various stores that had not been unloaded from the ships, including equip-ment belonging to the savants; aboard the flagship L'Orient there had been gold, silver and jewels looted from Malta during the voyage to Egypt. A French team of archaeologists has relocated the site of the wreck of L'Orient, and successfully recovered gold coins and a bronze plaque inscribed with its name, as well as recognizable parts of the vessel and remains of some of the crew.

The destruction of L'Orient and the other warships was the destruction of Napoleon's ambition of conquering the East. Although Napoleon put a brave face on it and still talked of leading an army to India, the French were now marooned in Egypt. They had transport ships, but no warships to protect them from the Brit-ish Navy, which threatened their supply route from France. Much more significant than the military outcome were the political con-sequences: with the French expedition so weakened, Turkey broke

off negotiations with France and joined the alliance of France's enemies, and eventually a Turkish army would be gathered to march
against the French in Egypt. The loss of the gold and silver aboard
the warships was also a blow, because rather than follow his usual
practice of allowing his soldiers to live off the land, taking whatever
they wanted, Napoleon was attempting to win over the native population by paying for everything the army needed. He was rapidly
running out of money.

The victory at Aboukir Bay was the greatest that had been
achieved in the war against France and the first major setback for
Napoleon, up to then hailed as invincible in French propaganda.
Nelson became a British national hero and was heralded as 'Lord
Nelson' in *The Times* of London even before King George III had
raised him to the peerage. Rather than the Battle of Aboukir Bay, it
became known as the Battle of the Nile and Nelson became Baron
Nelson of the Nile, with an annual pension of £2,000 for the rest of
his life. Among many other honours and gifts that Nelson received
was one of macabre practicality from Captain Hallowell of the *Swift-
sure*, one of the warships that had taken part in the sinking of
*L'Orient*, who presented Nelson with a coffin. The letter from Hallowell accompanying the gift began, 'My Lord, Herewith I send you
a Coffin made of part of L'Orient's Main mast, that when you are
tired of this life you may be buried in one of your own Trophies.' On
9 January 1806, this coffin, with the body of Lord Nelson inside,
was lowered into a vault in St. Paul's Cathedral in London.

For the savants, the Battle of the Nile meant the loss of most of
their reference books and many scientific instruments as well as the
hope of a swift return to France after their task of recording the
country was complete. At Alexandria, the initial reaction was fear
that the British fleet might now attack the port, and so the inventor
Nicolas Conté devised furnaces to provide red-hot cannon balls for
use against the enemy ships and a floating pump to deal with fires.
Conté later set up workshops in Cairo, where he and his assistants
set to work on replacements for much of the scientific and military

supplies that had been lost, the first problem being to produce the tools necessary for the manufacture of precision equipment. Among the scientific instruments manufactured by these workshops were compasses, microscopes, telescopes, surgical apparatus, and drawing and surveying equipment, while sword blades, bugles, cloth and even uniform buttons were made for the army.

On 22 August 1798 Napoleon set up the Egyptian Institute of Arts and Sciences at Cairo, selecting a committee of seven who were to elect further members. The Institute had four sections (Mathematics, Physics, Political Economy, and Literature and Arts) and included the most distinguished and most promising of the savants, with the mathematician Joseph Fourier appointed as its Perpetual Secretary. The Institute's achievements were of lasting importance, whose benefits remained long after the fighting in Egypt was forgotten and when the massive loss of life at the Battle of the Nile and Napoleon's campaigns in Egypt over the next two years had become merely footnotes in history. Napoleon attached great importance to the Institute and to the work of the savants in general, which was reflected in the accommodation provided for them in the suburb of Nasriya at Cairo, in a building complex centred around a former Mameluke palace. Here there were meeting rooms, a chemical laboratory, library, observatory, printing press, zoological and botanical gardens, areas for agricultural experiments, Conté's workshops, and even mineralogical and archaeological collections, as well as a small natural history museum. The savants lived in rooms in the palace and in some of the surrounding houses, holding formal meetings in what had once been the harem, while informal meetings took place every evening in the gardens.

The purpose of the Institute was wide-ranging: the research, study and publication of natural, industrial and historical facts about Egypt, and the propagation of the resulting knowledge. From early on the Institute adopted a multi-disciplinary approach to the problems presented to it, and in time it was responsible for the building of hospitals, irrigation schemes, sewage systems, quarantine stations to

combat infectious diseases, and a postal network, as well as projects
to study almost every aspect of the country. The varied work of the
Institute was published from 1809 to 1828 as the *Description de
l'Égypte* (Description of Egypt), much of which was devoted to the
antiquities of the country, a very important advance for Egyptology
(although Egyptology was not a term used until the mid-nineteenth
century), not least because in some cases it became the only record of
monuments that were destroyed in the years after the French left
Egypt. The engravings of hieroglyphs on the monuments that were
gradually published became a prime source of material for all the
would-be decipherers, but until those engravings were compared
with the monuments themselves, after the hieroglyphs had been
deciphered, nobody realized just how the inaccuracies and errors in
these engravings also led researchers astray.

Once established in the Nasriya district, the savants began to
enjoy themselves, and although each pursued his own speciality,
they found the unprecedented interaction with other experts exhilarating. They were fascinated by everything that was being
discovered about the country and gradually acclimatized to its customs. Several savants took to drinking Turkish coffee and smoking
the narghileh (water-cooled pipe), and they let their beards grow,
having discovered that a shaven chin was regarded as the mark of a
slave, but their enthusiasm seldom extended to Cairo itself. Most
soldiers and savants consistently agreed on their opinion of the city.
The engineer Villiers du Terrage, while admiring the beauty of
Cairo's 300 mosques, was well aware that the streets were quite dis-
gusting, and the artist Denon wrote dejectedly that he saw in Cairo
'a huge population, lengthy spaces to cross but not one beautiful
street, not one fine monument: a single vast square, but which has
the appearance of a field...palaces surrounded by walls, which
sadden the streets more than they embellish them; the living quarters of the poor, more slovenly than elsewhere.' Napoleon's view of
Cairo, with its 300,000 inhabitants, was of a city with 'the world's
ugliest rabble.'

While the French failed to warm to Cairo, many of its inhabitants did not like the French, who offended local customs and religious traditions in many ways: by forbidding the burial of the dead, however holy, in the interior of the city; by levying a building tax that required the examination of documents about buildings that were considered private and sometimes had a religious significance; and by many minor regulations, such as enforcing the sweeping of streets and clearing away rubbish. The public behaviour of the French was often regarded as immoral by the inhabitants, who could not understand why they made widespread use of female prostitutes and not young boys. Worse still, the men became seriously worried when their own wives and daughters began to imitate the freedom of European ways, appearing in public unveiled in the company of the invaders: when the French lost control of Egypt in 1801, many of these women were beheaded for their conduct. The ill-feeling towards the French was exploited by religious fanatics and by agents of the Mamelukes who promised that Ottoman armies were on their way to drive the French from Egypt and that Moslems should rise up in holy war against the French. Even though such calls for a holy war were shouted from the minarets of the mosques during prayers (five times a day), the French remained unaware of the situation, and the uprising of 21 October took them completely by surprise.

The revolt started early in the morning, with barricades being built in the streets, armed men gathering at the mosques and the shops closing down. The French troops were put on alert at eight o'clock, but the French administration still did not appreciate the danger, and Napoleon and three of his generals even left to inspect some fortifications being built outside the city. By ten o'clock the news that there was a widespread uprising reached Napoleon and he hurried back to find bodies in the streets, fires breaking out across the city, and the non-Moslem areas under attack. At General Caffarelli's house the mob killed four savants and looted or destroyed many scientific instruments.

Only the Citadel, the barracks, the army headquarters at Esbekiya Square and the buildings of the Egypt Institute remained in French hands. Having control of most of the city, the mob began looting the warehouses, regardless of whether they were owned by Christians or Moslems. The Egypt Institute, two miles from Esbekiya Square, became surrounded by a hostile crowd, and help only arrived in the evening, when a company of grenadiers brought forty muskets that few of the savants knew how to use. The mathematician Gaspard Monge organized the defence of the Institute, largely to protect the instruments and equipment stored there, but the night passed quietly and the next morning the savants held off their attackers for two hours with sporadic musket fire until two French army patrols came to their rescue. To restore order Napoleon concentrated on the centre of the insurrection, the El Azhar Mosque, which he bombarded with artillery. Three infantry battalions with bayonets fixed and 300 cavalry with drawn sabres then converged on the mosque and took it by storm. Several hundred rebels were captured, and the mosque was systematically looted and deliberately desecrated. By nightfall the fighting was over; the French had lost about 300 men while up to 5,000 inhabitants may have been killed.

Although the rebellion in Cairo was by far the worst, uprisings occurred in other places under French rule, all of which were quickly suppressed, but there was still a danger from the Mamelukes. After his defeat at the Battle of the Pyramids, Murad Bey had escaped and was gathering another army. On 25 August 1798, some eight weeks before the October uprisings, General Desaix had set out for Upper Egypt in search of Murad, taking with him an infantry force of 2,861 men and two field guns on a journey of over 3,000 miles that was to last many months. Although Desaix was a brilliant general, the campaign could not possibly succeed, due to Murad's ability to exploit the desert terrain to his advantage, but this epic journey was to have a lasting impact on Egyptology and the study of hieroglyphs, because for much of the time Desaix was accompanied by the artist Vivant Denon. Catching up with Desaix in early November, Denon journeyed up and down the Nile Valley for nine months with the forces commanded by General Desaix and by General Belliard, who had just arrived with reinforcements.

At age fifty-one, Dominique Vivant Denon was one of the oldest of the savants and had already enjoyed a distinguished career, having once been part of the Court of Louis XV and a favourite of Madame Pompadour. After studying art and literature, he had produced drawings, paintings and several books, including one of pornographic drawings, and written a successful play. He was an experienced traveller and had been a diplomat in Russia, Sweden, Switzerland and Italy. At the start of the French Revolution he was in Venice, and although he managed to return to France and have his name removed from the proscribed list, all his property had been confiscated, and he was forced to eke out a meagre living by selling his drawings and writing. Denon became one of the circle of artists and intellectuals that frequented the Paris salon of Josephine, where he met her husband Napoleon. Suspicious of Denon's connections with the pre-Revolutionary monarchy and considering him too old, Napoleon at first refused to include him in the Egypt expedition. It was Josephine who persuaded Napoleon to invite Denon, thereby setting in motion a sequence of events that would bring ancient Egypt to the attention of the western world and bring Egyptian influence to several decades of European styles and fashions.

Since landing in Egypt, Denon had spent every moment he could sketching all that he saw and had already built up a large collection of drawings. He worried about running out of pencils, since the savants were always short of supplies, and constantly requested more pencils from Conté, who manufactured them in his Cairo workshops. When supplies did not catch up with Desaix's army, Denon improvised pencils by melting lead bullets. The shortage of pencils was not so serious a problem for him as the shortage of time. For safety, he always had to travel with the army, which moved fast in pursuit of Murad Bey, seldom stopping long in any one place, and he usually only had a matter of minutes in which to make a complete sketch before being forced to move on once more. The army moved southwards up the Nile Valley, hemmed in by desert on either side. The disappointment of Egypt's ancient remains, with virtually nothing at Alexandria and only pyramids and the Sphinx near Cairo, was soon replaced by amazement at the incredible temples and tombs of Upper Egypt. Denon snatched every fleeting opportunity to explore their remains and make rapid sketches, soon becoming aware that almost all the ruins were covered in hieroglyphs. The limitations of his sketching were immediately apparent, and he accurately summed up the problem of the hieroglyphs: 'It would take months to read them, supposing the language was known: it would take years to copy them.' When the army reached Dendera, the sight of the temple so amazed the soldiers and officers alike that they spontaneously left the column and rushed to explore it. For once, Denon had the rest of the day for drawing, but like the soldiers he was overwhelmed by the magnificence of the structure and hardly knew what to draw first. Each facet of the architecture, every relief sculpture, every painting, the profuse hieroglyphic inscriptions that covered almost the whole surface of the temple, inside and out, all called for his attention at once. He recorded that: 'Pencil in hand, I passed from one object to another...I had not eyes or hands enough, and too small a head, to see, draw and classify everything that struck me. I felt ashamed of the inadequate drawings that I was making of such sublime things.' Totally absorbed, Denon sketched furiously until the light failed, and only then did he realize that the army had gone – all except his friend General Belliard who had stayed to keep a protective eye on him; they were obliged to gallop their horses to catch up with the army before night fell.

Dendera was the first major ancient site that the army had seen close up and it profoundly affected many of the soldiers. Built in a completely unknown style of architecture, the temple was covered in hieroglyphs, and on the ceiling of one room was carved a marvellous circular representation of the zodiac. Denon was astonished by the ruins and recorded: 'I would like to be able to excite in the reader's mind the sensation that I experienced. I was too much amazed.' That evening a junior officer approached him and expressed what many others felt: 'Since I came to Egypt, fooled by everything, I have been constantly depressed and ill: Dendera has cured me; what I saw today has repaid me for all my weariness; whatever may happen to me during the rest of this expedition, I shall congratulate myself all my life for having been in it.'

The army continued to chase after the Mameluke leader Murad Bey, and Denon continued to sketch whenever he could, generally 'on my knee, or standing, or even on horseback: I have never been able to complete a single one to my liking, because for a whole year I have not once found a table sufficiently steady for using a ruler,' but on occasions friendly soldiers supported his drawing board and shaded his work from the fierce sunlight. On 27 January 1799 the army rounded a bend and saw the panorama of ancient Thebes for the first time, and in sheer amazement the soldiers halted and burst into spontaneous applause. For Denon it meant yet more frustration, because the army could not stop and all he could do was gallop from temple to temple, to the necropolis and back again, with a cavalry escort, before riding hard to catch up with the army. Continuing south past other ancient sites, they travelled 250 miles in ten days, arriving at Aswan on 2 February: the Mamelukes had left just two days before. Here, General Belliard recorded in his diary that the awesome cataracts and huge desert seemed to mean that nature was saying 'Stop, go no further.' After two days at Aswan, they began the march back north, and over the next fifty days covered about 550 miles as they moved up and down the Nile searching for the Mamelukes.

Because the army passed some of the ancient sites several times, Denon was gradually able to amass a series of sketches of each site, and his folder of drawings became increasingly precious – it was never out of his sight and he used it as a pillow when he slept. He also collected whatever portable antiquities he could, such as pottery vessels, statuettes and even the mummified foot of a woman he found in a tomb in the Valley of the Kings. This foot later inspired the short story *Le Pied de Momie* (The Mummy's Foot) published in 1840 by Théophile Gautier – although not the first work of fiction to feature a mummy, it started a whole genre of romance and horror stories that would later generate a series of horror films. The pride of Denon's collection was a papyrus roll with hieroglyphic writing that he acquired in Thebes – while negotiating the surrender of some sheikhs, a mummy was brought to him with the papyrus roll clutched in its hand, and he was almost overcome with emotion: 'My voice failed me...I didn't know what to do with my treasure, so afraid of destroying it; I didn't dare touch this book, the most ancient of books known up to this day...Without thinking that the writing of my book was as unknown as the language in which it was written, I imagined for a moment that I held the *compendium* of Egyptian literature.'

While Denon did what he could to record and collect antiquities, the army sometimes managed to catch up with the Mamelukes and force a battle. Murad's tactics were always the same: arriving in an area a few days before the French, he would stir up the local peasants with his propaganda and then enlist or conscript them into his army. When the French attacked, Murad made sure that the peasants took the brunt of the battle, and while the French were occupied fighting them, the Mamelukes rode off into the desert. Although the peasants were killed in their thousands, the French inevitably tended to lose more men than the Mamelukes, who moved to another place in the Nile Valley to continue their war of attrition with a fresh contingent of peasants. From the French point of view, it seemed that this would continue until they had no army left, but there was growing dissension among the Mameluke beys themselves, who tended to run away as soon as they engaged the French in battle, hoping to preserve their own men while the

French killed the followers of their rivals. By mid-March 1799 Murad's forces began to disintegrate, leaving the French theoretically in control, but they could not give up the pursuit as the beys could combine again into a force large enough to overwhelm them. Desaix split his army many times to try to deal with the dispersed Mameluke bands, while a separate force under Belliard continued to march up and down the Nile Valley, fighting the Mamelukes wherever they could be found.

Also in March, a group of engineers led by Chief Engineer Pierre Girard was sent from Cairo to join General Belliard in order to study the River Nile and see how it could be used to increase the fertility of the land. Two of these engineers, Prosper Jollois and Édouard de Villiers du Terrage, were particularly impressed by the ancient ruins and were determined to record as much as possible, a resolve that was strengthened when on 25 May they met Vivant Denon at Qena, who showed them his drawings of the incredible ruins at Dendera. While stationed at Qena, on the opposite bank of the Nile, they made many visits to Dendera, producing measured plans, sections and perspectives, and studying the architecture and building methods. As engineers rather than artists, their approach to recording the monuments was scientific, and their record of the zodiac was far more precise than that achieved by Denon. Girard strongly disapproved of their interest in the ruins and did what he could to stop them, but they managed to complete their work on the hydrology of the Nile and still have time at the monuments, and were even supported against Girard by General Belliard, who had been convinced by Denon of the importance of their work. As well as Dendera, the two engineers visited the island of Philae near Aswan, the temples at Kom Ombo, Edfu and Esna, and the temples and tombs of Thebes, making plans and architectural drawings, as well as copies of hundreds of hieroglyphs. Like Denon before them, they ran out of the pencils made by Conté at Cairo and had to improvise by melting bullets and pouring the lead into hollow reeds.

On 19 July 1799 Napoleon set up two commissions of savants, led by the mathematicians Joseph Fourier and Louis Costaz, to make a scientific study and accurate record of the ancient monuments in Upper Egypt, but it was only when Denon returned to Cairo in the middle of August that they appreciated the scale of this task. As Denon began to relate all he had seen and to show them his drawings and artifacts, they were completely astounded and realized that the few monuments in Lower Egypt, including the pyramids, were nothing in comparison with the wonders of Dendera or Thebes. Indeed, Denon's evidence highlighted just how important were the hieroglyphs with which almost everything seemed to be covered. If they could decipher the hieroglyphs, they would be able to understand the monuments, but for now all they could do was copy them. In the journals written by the savants they only described the physical appearance of all these hieroglyphs, being unable to comment on their meaning. The two commissions left Cairo on 20 August, meeting up with Jollois and Villiers when they reached Upper Egypt. Fourier and Costaz wisely decided not to repeat the work of these two engineers, but to concentrate on what was still waiting to be recorded. The result was a mass of notes, drawings, papyri, mummies, statuettes and a whole range of other artifacts that were carried back to Cairo for further study.

On the same day that Napoleon set up the commissions to study Upper Egypt, one of the clues to deciphering the hieroglyphs was discovered at Rosetta. A party of soldiers was strengthening the defences of the dilapidated Fort Rachid, renamed Fort Julien by the French, a couple of miles north-west of Rosetta. A ruined wall was being demolished when a soldier called D'Hautpoul uncovered a damaged dark grey stone slab with inscriptions covering one side. The officer in charge, Lieutenant Pierre François Xavier Bouchard, thought that it could be of some importance and informed his superior, Michel-Ange Lancret. On examining it, Lancret found there were three inscriptions in three different scripts: one was in Greek, which he could recognize, one was in hieroglyphs, and one was in another unknown script. Once translated, the Greek text showed that it was a decree by the priests of Memphis, dated 27 March 196 B.C., commemorating Ptolemy V Epiphanes who ruled Egypt from 204 to 180 B.C. It was immediately assumed that the three inscriptions represented the identical text in three different scripts and languages, thereby providing the key to deciphering hieroglyphs.

The substantial stone, nearly four feet high and weighing three-quarters of a ton, was given to Bouchard to transport to Cairo, while Lancret, only recently elected a member of the Institute of Egypt, informed his colleagues that 'citizen Bouchard, officer of the engineers, has discovered in the town of Rosetta some inscriptions whose examination may offer much interest.' As soon as the stone was seen by the savants in Cairo they began devising ways of making accurate copies of the inscriptions, and eventually a variety of rubbings, drawings and casts were made. They were all excited at the prospect of a breakthrough in the problem of reading hiero-glyphs, but their euphoria was premature – it would take another twenty-three years of hard work and much bitterness before anyone could even begin to read the hieroglyphs. The savants began to refer to the Fort Rachid discovery as the Rosetta Stone (*Pierre de Rosette*), and it was soon to become one of the most famous monuments in the world.

While the savants were at last seeing spectacular results to reward their hard work and suffering, and the forces of Generals Desaix and Belliard had restored some order and control in the Nile Valley, the bulk of the military expedition led by Napoleon was not having much success. The British blockade of Egypt ensured that very little in the way of supplies, people or information could pass through, but more seriously the Turks were openly hostile, and in February 1799 Napoleon had learned that they were planning a two-pronged attack with both a seaborne army and another army marching south to Egypt through Syria. To counteract this threat, Napoleon took the majority of his own army on an expedition into Syria in an attempt to destroy the Turkish land forces before they were fully organized. If his plan had succeeded and if the Mamelukes and Arabs had joined his army, Napoleon might then have achieved his dream of emulating Alexander the Great and marching on India. For once Napoleon was unlucky, the Syrian campaign was a disaster, and he was forced to retreat from Acre, having merely slowed but not stopped the Turkish advance towards Egypt.

Arriving back in Cairo on 14 June, Napoleon staged a triumph as if he had been successful, but it was obvious that it was only a matter of time before the French lost control of Egypt, unless they received supplies and reinforcements from France. In July news came that a Turkish fleet accompanied by British ships was disembarking a huge army at Aboukir (where Nelson had destroyed the French fleet several months earlier). Marching rapidly to Aboukir, the French decisively defeated the Turks. While exchanging prisoners with the British commander, Napoleon learned that France was in a severe economic crisis, Royalists were looking to restore the monarchy and a coup d'état to replace the Directory was increasingly likely. Although this was the chance of power that Napoleon had been waiting for, it looked as if he could not return to Paris in time. On 17 August he hurriedly left for Alexandria and sailed for France five days later, taking with him Generals Berthier, Lannes and Murat, a small group of military aides and bodyguards, the civilian savants Monge, Berthollet and Denon, and among the soldiers another three members of the Institute of Egypt. At the last minute the poet Parseval Grandmaison, who had learned of this secret departure, jumped on board one of the ships, clung to the rigging and begged to be allowed to go back to France; only the intervention of the other savants prevented Napoleon having him thrown off the ship.

Lucky to reach France just after the news of his recent defeat of Turkish forces at Aboukir, Napoleon was able to present his Egypt campaign in a favourable light. After nearly a month of feverish intrigue with his supporters, he managed to take control in a coup d'état on 9 November, so that in place of the Directory, France was now governed by a triumvirate of consuls, with Napoleon appointed First Consul for a period of ten years. Later he was to drop even this semblance of democracy, ousting the other two consuls and crowning himself emperor in December 1804.

Once Napoleon had left Egypt, overall command fell to General Kléber, who had deliberately not been warned of Napoleon's departure and was furious at his abandonment of the expedition. Ignoring Napoleon's written instructions (prefaced by: 'when you read this, Citizen General, I will be in the middle of the vast sea'), Kléber immediately began negotiations with the British commanders to evacuate the French from Egypt, and an agreement was reached and a treaty signed. On 4 February 1800 about forty savants prepared to leave. Initially delayed by an outbreak of the plague, they eventually boarded ship at Alexandria on 27 March, taking all their collections, including the Rosetta Stone, and were later joined by yet more savants. If they had sailed for France at this point, the Rosetta Stone would now be on display in the Louvre rather than in the British Museum, but when news of the treaty reached England it was repudiated by the British government, who insisted on an unconditional surrender by the French, and so the expected permission for the ship to leave was not forthcoming.

After spending a month on board ship hoping daily to set sail for France, the savants realized that there was no longer any chance of an immediate evacuation, and so they disembarked. Bitterly disappointed, and weary of the hardships of Egypt, they reluctantly went back to their work. It was to be another eighteen months before they could leave, and the continuing negotiations with the British even covered what items the savants could take with them: at times the bargaining bordered on violence. Eventually the savants were allowed to keep all their records and the majority of their collections, but the British took the most significant items – including the precious Rosetta Stone.

Soon after its discovery in 1799, ink rubbings of the three inscriptions on the Rosetta Stone had been sent from Egypt to scholars

across Europe, including those at the Institute of France in Paris.
Following capitulation to the British, the antiquities that had been
surrendered (fifty tons in all) were shipped to England, and in February 1802 the Rosetta Stone arrived in Portsmouth on board the
appropriately named HMS *L'Égyptienne*, a frigate taken from the
French in Alexandria harbour. The Rosetta Stone was transported
to the Society of Antiquaries of London, where plaster casts were
made for the Universities of Oxford, Cambridge, Edinburgh and
Trinity College, Dublin, and engravings were undertaken for distri-
bution to the major academic institutions throughout Europe –
even to the National Library at Paris during the brief cessation of
hostilities between Britain and France from March 1801 to May
1803. The stone itself was finally deposited in the British Museum
at the end of 1802, but engravings of the Greek, demotic and hieroglyphic texts were not published by the Society of Antiquaries
until 1815.

Because Napoleon manipulated the official figures, the precise
number of French casualties from his campaign is unknown, but of
the 50,000 troops that he took to Egypt, over half died there and
several thousand more were blinded or crippled. Their three-year
adventure took its toll on the savants too; at least twenty-five of
them died in Egypt (mostly from plague and other diseases,
although some were killed in battle and even assassinated), and all
of them were ill at some point. Disease was a constant hazard for
soldiers and savants alike, and in an era before sunglasses almost
everyone suffered from ophthalmia, an infection of the eyes caused
by the irritations of strong sunlight and blown dust and sand. The
worst cases of ophthalmia caused permanent blindness: one in five
Egyptians had at least one eye missing, and many savants spent
weeks or even months unable to work properly because of the condition of their eyes. Soldiers and savants also suffered from leeches
in their throats, stomachs and nasal passages through drinking
impure water; dysentery was endemic, and they frequently had
sunstroke and various fevers, including typhoid. On top of all these

ailments and frequent shortages of food and water, there was an ever-present threat of bubonic plague, from which some savants and hundreds of soldiers died.

From a military point of view, the expedition was a failure. At the end of three years of fighting the French had not gained an eastern empire, but had finally been forced to abandon Egypt. The one success of the campaign was that it broke the stranglehold which the Mamelukes had on the country. Ottoman rule was restored in Egypt, and a decade later the Mamelukes were wiped out. After the expedition it gradually became safer to travel up the Nile, and Europeans soon began to explore the region. For Napoleon's political ambitions, it had been a close-run thing, but he had succeeded in returning to France in time to seize power. The real success, though, belonged to the savants, whose return to France with their records and their collections was a scientific triumph at least as great as the military side of the expedition was a failure. There would always remain a special link between France and Egypt.

Travelling with Napoleon in 1799, Vivant Denon had been one of the first savants to return to France. He immediately began to sort out his notes and have plates engraved from his sketches. The resulting illustrated book, *Voyage dans la Basse et la Haute Égypte* (Journey in Lower and Upper Egypt), was a huge success when it was published in 1802. During the nineteenth century over forty editions of the book appeared and it was translated into several languages. More than anything else, this huge bestseller brought Egypt to the attention of the general public, not just in France, but throughout western Europe, and it was later emulated by others who returned from Egypt and wrote accounts of their adventures.

The official record of the work of the savants took somewhat longer to produce – the first two editors (Nicolas Conté and his successor Michel-Ange Lancret) died before the first volume was ever published. The third editor, the engineer and geographer Edme-François Jomard, took two decades to complete the work, which was published in Paris as the *Description de l'Égypte* in twenty monumental volumes from 1809 to 1828, including a preface on the history of the country by Joseph Fourier. With its hundreds of illustrations, many of them coloured, this work probably had as much impact on the scholars of western Europe as Denon's book had on the general public, but because of its size it was very expensive and relatively few copies were produced.

The publications that resulted from Napoleon's expedition to Egypt presented readers with things they had never seen or heard of – remains of a civilization thousands of years old, whose very existence was only previously known to a few scholars. As Denon put it, Egypt was 'a country which Europe scarcely knows but by name.' Suddenly the world as viewed from western Europe was much larger, older and stranger than it had been before. In itself, this was enough to spark a wave of Egyptomania across western Europe, but in France Egyptian themes were actively promoted by Napoleon, who was conscious of the need to create a style that would lend dignity to his empire and court without employing any of the symbolism that had been used by the preceding regimes. Having gained political power, Napoleon had been able to portray the military fiasco in Egypt as a triumph, and Egyptian motifs in architecture, furnishings and ornaments became extremely popular. Six out of fifteen new fountains designed for Paris by a decree of 1806 were in an Egyptian style, and sphinxes, pylons and pyramids were often used as architectural ornaments. Such use of Egyptian styles brought them to the attention of a wider public, as did the use of Egyptian settings in stage designs for plays and operas, the most influential being the sets for Mozart's *The Magic Flute*, which was performed in opera houses across Europe.

The plates in Denon's book were used as a source of inspiration by architects and designers, and particularly influenced the manufacture of furniture and most other items of interior design, at a time when Paris was the centre of European fashion. The porcelain factory at Sèvres, for instance, produced an Egyptian Service decorated with views of Egypt taken from Denon's drawings and completed with a magnificent table centrepiece that included models of a temple at Philae, obelisks from Luxor, the pylons from the temple at Edfu and avenues of ram-headed sphinxes.

In his attempt to make a clean break from the old French Royal dynasties, Napoleon even replaced the fleur-de-lys (lily) symbol (associated with the Bourbon monarchy) with the bee – a symbol taken directly from hieroglyphs. Although it was not known at the time, the bee was the hieroglyphic symbol for Lower Egypt – the symbol for Upper Egypt was the sedge plant, and pharaohs ruling the whole of Egypt were hailed as ⳤ⳥ (he of the sedge and the bee). Napoleon probably adopted the bee symbol on the strength of the statement by the ancient Roman writer Ammianus Marcellinus (who lived around A.D. 325–395) that the bee was the Egyptian symbol of royalty because 'in a ruler sweetness should be combined with a sting as well.' The bee became the predominant emblem of the Napoleonic Empire, but other emblems such as a star and laurel leaves were also used. The laurel leaves were an allusion to the Classical emblem of the victor, and Roman emperors had used the crown of laurels as a badge of office. The star was a five-pointed one, sometimes of the same design as the hieroglyph ✷ that was erroneously thought to signify 'divine' (but did actually mean 'star'); as such it was sometimes used with the bee in Napoleonic images to imply the meaning 'divine king.'

The Egyptomania which engulfed Europe highlighted the need to decipher the hieroglyphs. Many scholars, seeing a chance to acquire lasting fame and possibly fortune, embarked on a quest to find the key to these symbols – a quest that was to cause rivalry and bitter recrimination for many years to come.

# CHAPTER TWO

# (The Pupil)

When hieroglyphs were finally deciphered by Jean-François Champollion, he became a legendary figure in France: so much so that people began to talk about miraculous events in his childhood that foretold his brilliant success, and it is now difficult to distinguish between fact, fiction and exaggeration in the records of his early life. His father, Jacques Champollion, had been an itinerant bookseller from Valbonnais, to the south of Grenoble in the Alps, where life was so harsh that young and old men alike would leave their families for months on end to make a living as pedlars or even beggars. In 1770 Jacques settled in the town of Figeac in the Quercy region of southwest France, on the western edge of the Auvergne, where he set up its first bookshop in the market square, selling a variety of new and second-hand works including religious tracts, prayer books, political books and pamphlets, dictionaries, newspapers and practical books on subjects such as medicine and agriculture. On one of the pilgrimage routes to the shrine of Santiago de Compostela in Spain, Figeac provided Jacques with a good living at first, and two years later he bought a house there. In 1773 he married Jeanne-Françoise Gualieu, who came from a local bourgeois family engaged in some kind of manufacturing, but, surprisingly for the wife of a bookseller, she never learned to read or write. Both she and Jacques were thirty years old when they married, and in all they were to have seven children, although two sons died: Guillaume at birth and Jean-Baptiste when nearly two years old. Jean-François was the last child of a family of one other surviving son (Jacques-Joseph) and three daughters (Thérèse, Pétronille and Marie-Jeanne).

In January of the year in which Jean-François Champollion was to be born, his mother was very ill and almost paralysed with rheumatic pains – doctors could do nothing more for her. A local healer called Jacqou the Sorcerer was consulted, and having been treated with herbal remedies she started to recover. Jacqou added to his triumph by prophesying that she would make a complete recovery and that, despite the gap of eight years since her last child, she would give birth to a son who would be famous for centuries to come. As her recovery continued it began to be seen that his first prediction was true, and the whole town was excited by the prospect that his second prediction might come true as well.

Such, at least, is the local legend of the events leading up to the birth of Champollion, and it is easy to dismiss it as fiction, but, prophesies aside, the story is not totally improbable. In an era when physicians used leeches as frequently as modern doctors prescribe aspirin, herbal medicines from a 'sorcerer' may have been more effective, and it was certainly remarkable that in an age of minimal health care a forty-six-year-old woman should give birth to a healthy son less than a year after a severe illness. Jean-François Champollion was born in the family home at Figeac in the early hours of 23 December 1790, during the darkest hours of the year, and was baptized the same day in the Medieval hilltop church of Notre-Dame-du-Puy which still overlooks the town: his godfather was his twelve-year-old brother Jacques-Joseph, and his godmother was his aunt Dorothée Gualieu.

Champollion was a child of the French Revolution. The house in which he was born and spent his infancy, in the dark narrow rue de la Boudousquerie, was barely thirty yards from the small square where the guillotine was set up and where the Tree of Liberty, a symbol of the Revolution and a focus of political meetings and celebrations, was planted. Among the earliest sounds he heard were the cries of the crowd at the terrifying executions and the rowdy revels of Revolutionaries in the square. Arising from the growing tension between the privileged and largely untaxed nobility and the rest of the population who were overtaxed and increasingly oppressed, the Revolution had erupted in 1789, the year before he was born. In the next ten years the Bourbon dynasty was removed from the throne of France, and King Louis XVI and many of the nobility were executed by guillotine. The Catholic religion was suppressed and the monarchy was replaced by a succession of improvised governments.

During the Revolution people frequently attacked their enemies by denouncing them to the authorities for being in some way opposed to the new regime, and many were executed merely because they were alleged to have spoken against the current Revolutionary government. In Figeac, as elsewhere in France, it was not safe for children to play in the streets and the schools were closed down because most had been run by the religious orders. Largely confined to the house, Champollion was robbed of his childhood, but in some ways he never was a child. Like many gifted children, he quickly outgrew his contemporaries and was more at ease in the company of adults. Until he was seven years old, in the year that Napoleon's expedition set out for Egypt, Champollion received no education. His mother seems to have been ill yet again and was incapable of looking after him, his father was often absent, and it was probably left to his brother and sisters to take care of him. For much of the time he was confined to the house or the bookshop and left to amuse himself. For a family of five children the house was small and narrow: completely surrounded by tall buildings, it had a ground floor, two upper floors and a *soleilho* above. A feature of houses in the area, *soleilhos* are open upper storeys with roofs supported on columns of stone or wood. In a crowded town the *soleilhos* performed the function of backyards, used for storing firewood, drying washing and even growing plants, and doubtless children used them as places to play.

Champollion's high intelligence meant that he was easily bored, and his limited horizons made matters worse, often resulting in violent mood swings – one moment he would be engaged in a boisterous game, the next he would be lost in thought or quietly studying something that had attracted his attention. He developed a strong character that would later help him survive the problems brought about by his impatience and a quick temper that also emerged at this time. His favourite animals were lions, and he called himself 'lion,' probably as an abbreviation of Champollion that was easier to pronounce – later, when a student in Paris, he was to sign some of his letters with the Arabic phrase *Assad Saïd al-Mansour* (Victorious Lion).

Surrounded by books, but given no formal teaching, he tried to teach himself how to read and write, apparently with some success, which legend sees as the beginning of a passion for deciphering unknown scripts. He also began to copy drawings (indeed, to Champollion 'writing' meant copying words by drawing), which does seem to have been the beginning of a talent and enthusiasm for drawing that lasted all his life. The flexibility of mind that allowed him to see a script as a collection of drawings and a collection of drawings as a script probably stems from this time. One other lifelong trait showed itself early, since he was often to be found huddled close to the fire – like Napoleon, Champollion loved warmth and hated to be cold. The fireplace in the family home had a worn escutcheon on the stone lintel that probably represented two leaping dogs flanking a tree. In the eyes of Champollion, these animals were lions, reinforcing his attachment to his adopted symbol.

When Champollion was only four, his sixteen-year-old brother Jacques-Joseph started his first job in the municipal offices at Figeac. Although he only had a basic education, cut short by the Revolution, Jacques-Joseph learned quickly and was ambitious, and soon he was in charge of the registration of new laws and passports – documents needed by anyone travelling within France, not just to foreign countries. He already had a love of books and an interest in ancient history which would run parallel to that of his younger brother, but like so many other young men at that time, he was also excited by the conquests made by the 'invincible' General Napoleon. Early in 1798 Jacques-Joseph heard that his hero Napoleon was organizing an expedition and tried to join it: he was doubly disappointed at not being accepted into the army when he subsequently found out that the destination was Egypt. A few months later in July, Jacques-Joseph's father arranged a job for him with his cousins in distant Grenoble within Chatel, Champollion and Rif, a small import and export business specializing in textiles which nevertheless conducted business as far away as America. By this time Jacques-Joseph had managed to improve his education, showing talent in Latin, Greek and ancient history, and he continued to study in his spare time.

As he had begun to teach his younger brother in the months before moving to Grenoble, Jacques-Joseph's departure deprived Champollion of both his beloved brother and his education. In November 1798, just before his eighth birthday, he started to go to the recently reopened local primary school for boys at Figeac, but was unhappy there, not being accustomed to discipline and disliking many of the lessons. Some subjects were too easy, and the mechanical methods of teaching bored him, while others failed to interest him at all. He had a complete aversion to mathematics, particularly mental arithmetic, and was constantly in trouble for this and for his idiosyncratic spelling. It soon became obvious that he was not suited to the school any more than the school was suited to the special needs of this gifted child, and at the instigation of his brother he was taken away and put in the care of Dom Calmels, a private tutor.

Once a Benedictine monk (before the Revolution closed the monasteries), Dom Calmels taught Champollion for two years. His manner of teaching included walks in Figeac and the surrounding countryside discussing anything and everything that his pupil noticed or questioned, in order to develop the boy's skills in observation and reasoning. At that time Figeac was a delta-shaped cluster of mainly Medieval sandstone buildings, crowded within Medieval town walls and dissected by narrow lanes. Set on the gently rising slope of the north bank of the River Célé, views of the rolling green hills surrounding the town could only be glimpsed from the few open spaces.

Champollion made good progress in learning Latin and Greek, displaying an early gift for linguistics, and his excursions around the town stimulated an interest in natural history as well as art and architecture, but overall his progress was erratic. His spelling remained weak, and he was still subject to sudden changes in mood. After a year of teaching Champollion, Dom Calmels wrote to Jacques-Joseph about his progress, admitting that: 'he has a great appetite, a great desire to learn; but this appetite and desire are drowned in an apathy, a carelessness which it is difficult to deal with. There are days when he appears to want to learn everything, others when he would do nothing.' It seems that at this time it was only his love and respect for his older brother that drove Champollion, and Jacques-Joseph began a long process of encouragement, coercion and sometimes bullying in his attempts to motivate and guide him.

Gradually Champollion outgrew his tutor's ability to teach him, and Dom Calmels advised that a teacher in Grenoble might be better able to develop his apparent genius. Consequently, Jacques-Joseph wrote to his brother, warning him that: 'If you wish to come and stay with me, it is necessary that you learn something quickly; the ignorant are good for nothing.' In March 1801 Champollion was sent over 200 miles away to live with his brother in Grenoble – his mother would never see him again. Even at a distance, Jacques-Joseph had been the most influential person guiding his education, but now he took complete control. The boy arriving in Grenoble was ten years and three months old, with a swarthy complexion, black curly hair and very dark almond-shaped eyes whose sparkle betrayed a lively intelligence and a fiery temper. As he stepped from the coach, the town that he saw was prosperous and prestigious. Grenoble, in south-east France, is set in a cradle of mountains, and from the Grande Rue (Main Street), where Jacques-Joseph lived and worked in his cousins' business, it was possible to look up any street and see a mountain – some capped with snow all year round. The small town was largely confined to the area enclosed by the seventeenth-century fortifications (now mostly demolished) on the flat plain by the confluence of the Rivers Isère and Drac, which were often fast-flowing and ferocious, and responsible for numerous floods. Far more than his native Figeac, Grenoble was home to Champollion and he came to love it.

Part of the charm of Grenoble was that it was largely free from the memories of savagery and bloodshed that had blighted the recent history of many French towns. Because the nobility in the area had been in the forefront of the Revolution, forcing the King to reconvene the French parliament (the States General) ten years before any fighting broke out, much of the extremism and civil strife that had taken place elsewhere was avoided. Only at the height of the Reign of Terror was the stability of Grenoble threatened when it was learned that one of the Revolutionary Commissions, which were responsible for much of the bloody persecution of innocent victims, was about to be set up in the town. In July 1794 the leader of the Revolutionary government during the Reign of Terror, Maximilien Robespierre, was arrested and executed, and the collapse of his regime saved the lives of hundreds of people awaiting execution in the prisons of Paris, including that of a young aristocratic widow called Marie-Joseph-Rose de Beauharnais. It also stopped the installation of the Revolutionary Commission in Grenoble and protected the town from the witch-hunts and massacres that had taken place in the French capital. The crisis had passed, and in the years that followed Grenoble grew into the commercial and intellectual centre where Champollion would spend his adolescence, while the young widow Madame de Beauharnais remarried and became Josephine Bonaparte, future Empress of France.

Some time after Jacques-Joseph had arrived in Grenoble, he changed his surname from Champollion to Champollion-Figeac, which has sometimes been interpreted as a means of distinguishing between himself and his brother Jean-François: an early recognition that Jean-François Champollion was destined for great things. It was more likely a way to distinguish himself from his Champollion cousins in Grenoble, and it was also a mark of an ambitious bourgeois. Among those with social aspirations, a common practice was to give themselves grandiose names by attaching their place of origin to their surname. Before the Revolution, Jacques-Joseph would probably have changed his name to 'Champollion de Figeac,' to give it an aristocratic ring, but even after the Revolution it was still dangerous to pretend to be an aristocrat – like many of the nobility, Madame de Beauharnais had dropped the 'de' and signed herself 'Citizeness Beauharnais.' Although Jean-François was later sometimes referred to by the surname Champollion-Figeac, he appears to have rejected the fashionable name change, preferring to sign himself 'Champollion le Jeune' (Champollion the Younger). From his earliest childhood he was aware of the power of words and the importance of precision in conveying a message (despite his poor spelling), and later he was to use words, particularly nicknames, as others might use gifts or edged weapons. His conscious rejection of the surname his brother had adopted shows his different attitude to society, even at this early date. Champollion was never to have the social ambitions of his brother, nor, to his cost, the tact and diplomacy that would smooth the path of Jacques-Joseph through the turbulent times ahead.

During the first twenty months of his stay in Grenoble, Champollion was initially taught by a private tutor and then by his brother alone. He was still subject to changes in mood, and Jacques-Joseph regretfully wrote to Dom Calmels: 'At times enthusiastic and hurried, he seems to fear finding limits to his passion for learning, at other times he is weak and demoralized and everything seems to him to be an obstacle to overcome, a difficulty to resolve' –

comments strikingly similar to the report that Dom Calmels himself had given earlier. Jacques-Joseph blamed these characteristics on Champollion's mind, which was incapable of applying itself properly to any one task – a failing that would be remedied very rapidly. In November 1802 Champollion started to attend the private school of Abbé Dussert, who had a very good reputation as a teacher which was matched by the high price of his services: it was a great sacrifice on the part of Jacques-Joseph to pay for his brother to attend the best school – it was also emphatic proof of Champollion's prodigious linguistic talent that the person who knew him best would sacrifice so much to give this talent the greatest chance to develop.

Dussert's school operated in conjunction with the 'central school,' one of the state-run schools in France at that time. Champollion was taught languages by Abbé Dussert, but for most other subjects he attended the central school. His linguistic gifts became immediately obvious, and after a year he had made so much progress with Latin and Greek that he was allowed to begin to study Hebrew and three other Semitic languages: Arabic, Syriac and Chaldean. These languages were chosen to help the research that he had begun to do in his spare time. Although only twelve, he was fascinated by the origins of mankind, and since the Bible was the earliest known historical text which apparently described the creation of the world, he wanted to get past the errors in later translations and read it in the original language. He had already started to compile a chronology of ancient peoples and was encouraged in this by his brother. Jacques-Joseph was collecting quantities of books, benefiting from their sale at very low prices by people who had been ruined by the Revolution, and his growing library greatly helped his brother in his studies. This was the real starting point of a lifetime of research for Champollion.

At first Champollion seems to have been happy in Grenoble, not only immersing himself in ancient languages, but also taking pleasure in many of the lessons at the central school. Mathematics remained his weakest subject, which he still hated, but his drawing skills continued to improve and he developed a great interest in botany, even going on expeditions into the surrounding mountains to study and collect plants, just as he had once roamed Figeac with his tutor, studying and discussing everything they encountered. Early on in his time at Grenoble, he also had the incredible opportunity of meeting and talking with the newly appointed Prefect. This Prefect was Jean-Baptiste-Joseph Fourier, one of the foremost savants who had taken part in Napoleon's Egypt expedition and who had played a leading role in the administration of the Institute of Egypt at Cairo. Even before the expedition, he had made his name as a scientist and mathematician, and on his return to France at the age of thirty-three he was put in charge of writing the historical introduction or preface to the *Description de l'Égypte*. By early 1802, a few months after his return from Egypt, when he had taken up his new post of Prefect of the Department of the Isère, based in Grenoble, Fourier was something of a celebrity.

The savants who had been to Egypt called themselves 'Egyptians' and they were still fascinated by everything connected with that country. Fourier was no exception, and as well as his work on the *Description de l'Égypte*, he was immersed in studying various aspects of ancient Egypt, particularly zodiacs, whenever his duties as Prefect permitted. One of those duties was to inspect the government schools, and at one school he was so impressed with the enthusiasm for Egypt shown by a pupil called Jean-François Champollion that he invited him to see his collection of antiquities. Called to the Prefecture and faced with Fourier, famous scientist, 'Egyptian' and the most powerful man in Grenoble, the eleven-year-old was so nervous he could not speak, even to answer the questions Fourier put to him. Only when the Prefect began to talk of Egypt and show him some of the antiquities did he regain his composure. Seeing hieroglyphic inscriptions on stone and on fragments of papyrus, none of which could be understood, Champollion left the meeting not only determined to study and decipher the ancient script, but he also declared his conviction that he would succeed. Whether this was a premonition or just a bold statement of youthful enthusiasm, from this time on Champollion and Fourier were to influence each other until Fourier's death in 1830, and ultimately both were to lie virtually side-by-side in the Père Lachaise cemetery in Paris.

As Fourier was to find in Grenoble, Egypt was still a major talking point throughout France, since state propaganda had presented Napoleon's expedition to Egypt as an overwhelming success, Egyptomania was taking hold, and full backing was being given to the production and publication of the *Description de l'Égypte*. While the savants began collating their notes in Paris for the monumental publication, in Grenoble Fourier was working on its preface and Champollion was dividing his time between private tuition by Abbé Dussert and lessons at the state-run central school, enjoying himself and gradually becoming more settled. Unfortunately, the organization of state schools was about to be drastically changed, and it would be some time before he was again so content in his studies.

As the last of the Revolutionary governments, the Directory had been overthrown by Napoleon in 1799 on his return from Egypt and replaced by a triumvirate of Consuls – holding the position of First Consul, Napoleon was in effect a dictator. For several years he had been personally concerned with reorganizing the education system in France, and a law of 1802 provided for forty-five *lycées* throughout France, with the government paying the boarding fees for 180 pupils at each *lycée*. The *lycées* were designed to be elite boys' schools with a curriculum, uniform and military discipline set by the government. Even the 526 books in the library of each *lycée* were stipulated by the government to ensure that all these schools used identical texts. Of the books listed, 56 were French literature while 142 were works of Classical antiquity, and the syllabus was based on Latin and Greek literature and on mathematics, with natural history, chemistry, technical drawing and geography as additional subjects. Philosophy was deliberately omitted from the syllabus as a concession to the newly rehabilitated Catholic Church, and little time was given to history, which was regarded as a controversial subject.

The *lycées* were boarding schools and all the boys wore a uniform with a two-pointed hat. The uniforms were originally dark blue, but because the dyes were imported from the French colonies and became harder to obtain as a blockade of French shipping by France's enemies became increasingly effective, the government changed the regulation colour to grey. The boys were subject to military discipline, being divided into companies and ranks corresponding to those of an infantry regiment, and a military drum roll marked each change of timetable in their daily routine, from five-thirty in the morning to nine o'clock at night. In Grenoble the *lycée* was not set up until 1804, taking over the premises of the former central school. Early in that year Champollion sat and passed the examination which won him admission to the *lycée* and a bursary for his boarding fees, but he was not looking forward to taking his place in the new school, which he had already despondently labelled 'the prison.'

Within Continental Europe there had been an uneasy quiet rather than a proper peace, while the major powers adjacent to the French Empire, such as Austria and Russia, bided their time, but by May 1803 hostilities had again broken out between Britain and France. By the end of the following year Napoleon was at the height of his power, crowning himself Emperor and his wife Josephine as Empress on 2 December 1804. At the same time Champollion, nearly fourteen years old, felt utterly powerless: he had been at the *lycée* for two weeks and was seriously thinking of running away. Filled with bitterness and resentment against what he saw as an unnecessarily restrictive and time-wasting military regime at the *lycée*, he longed for the freedom to study as he wished and he missed the company of his former school friends – above all he missed his brother.

Champollion remained a boarder at the *lycée* for nearly two and a half years, during which time he wrote almost daily letters to his brother. These give a great deal of detail about his life, thoughts and feelings at that time, but almost all are undated, with very few clues indicating when they were written – a fact criticized by Jacques-Joseph: 'I have received, my very dear brother, the letter which you have written me, I can't tell you on what day, because, according to your custom, it has no date.' Jacques-Joseph was also irritated by the careless way Champollion invariably dashed off his letters and later warned him: 'I would rather you wrote your letters with more care, from the point of view of language; often you are incomprehensible and this is a great evil because one gets used to writing badly.' From the first, Champollion's formal writing seems generally to have been of a high standard, but it would be some years before his letters were more than rapid and sometimes garbled messages.

Resigned to the impracticality of running away, he frequently begged Jacques-Joseph to move him to another school, even though he knew that his brother wanted him to stay: 'Would you not be able to withdraw me from the *lycée*? I have been forcing myself not to displease you up to now, but it has become quite unbearable for me. I feel that I am not made for living squeezed together like we are here...If I stay here for long, I promise you I will die.' While some of his anguish was doubtless due to the stress of starting at a new school and living away from home for the first time – the kind of stress suffered by many boarding-school pupils – Champollion did have real problems with the regime at the *lycée*.

Locked into a rigid curriculum where he found some of the lessons far too easy and boring and others, like mathematics, too difficult and without interest, Champollion reverted to the violent mood swings that had blighted his earliest school days, sometimes being full of enthusiasm and sometimes depressed and apathetic. Even when he was enthusiastic, his independent mind and obvious abilities made enemies of some teachers who merely categorized him as lazy, insolent and rebellious. The irritations of the petty officials who ruled the *lycée* were not the only problem, for his bursary did not pay for much more than three-quarters of his expenses, the other quarter being paid by Jacques-Joseph, who could spare little else for his brother. Champollion was usually broke, and he felt the lack of resources more keenly because many of his companions came from wealthy families and were provided with more than adequate pocket money, while he lacked money for shoes, clothes and above all books. His poverty is a recurring theme in his letters to his brother at this time, and unfortunately it was to be a recurring theme throughout his life.

Worse than the lessons he was made to endure was the fact that he was forbidden to carry on his own studies during his leisure hours. He was forced to hide his books on subjects outside the curriculum, such as those on Hebrew and Arabic, and read them furtively at night after the watchman had done his rounds. The secret studies that he carried on each night not only left him exhausted, but also affected his eyes, particularly the left one, as he lay on that side in his bed to take advantage of the only light, from a nearby lantern in the street. Champollion was constantly torn by bouts of self-pity, and he frequently sought escape from his misery in the pleasures of his own research: 'The Oriental languages, my favourite passion, I can only work on them once a day...Greek, Hebrew and its dialects, and Arabic, here is what I burn and desire to learn.' His letters to his brother are littered with requests and demands for books, displaying scholarship and maturity beyond his years: 'Do not forget the rest of the books: add to it I beg you the dictionary and the Ethiopian grammar of Ludolph. Rest assured that I will not use it to the detriment of my other studies'; 'I beg you my very dear brother to send me a *Homer* of some kind: it will bring me a lot of pleasure if it is possible to send it to me this evening at the latest, I have an extreme need of it.'

Even in letters where his demands seem most brusque he frequently expressed his love and gratitude to his brother: 'Do you believe that I forget for a single instant everything that your affection has done for me and the paternal care that you have had for me for as long as I can remember?' He was later to repay this affection by his dogged support of his brother in desperate circumstances and in the long hours he spent teaching Jacques-Joseph's children. Champollion was also acutely aware that although his brother could not always afford to provide everything he needed, his parents were in an even worse financial position. When his father wrote to him offering help, Champollion, sensitive to the situation, tactfully refused: 'I don't need anything. I thank you for your kind offer. My brother provides all my needs. Be the bearer of my gratitude to him. I hope, by making the most of the advantages which I owe to his fraternal love, to prove to him that he has not helped an ungrateful person.'

As he progressed through his teenage years, in an alternating cycle of long bleak wintry months spent at the *lycée* and fleeting weeks of freedom and sunshine living with Jacques-Joseph during the school holidays, Champollion grew passionate about all forms of freedom, but particularly freedom of thought. With a love and respect for his brother not far short of hero-worship, and knowing that the *lycée* was the best education that he could afford to provide, only his gratitude for Jacques-Joseph's support prevented him from leaving, and yet the fact that he hated the *lycée* and felt oppressed by it did not mean that he was unsuccessful. He seems to have been popular with many of his fellow pupils and was the leading light of several of the learned societies that the inmates of the *lycée* were encouraged to set up. He was elected a corporal many times by his classmates, a post similar to a school prefect that was held for fifteen days at a time, and although he hated many of the lessons, he was still the best pupil in the school.

The restrictions that the *lycée* imposed on his own research were all the more galling to Champollion because during the holidays of the summer of 1804, before he took up his place at the school, he had composed what he later called his 'first folly,' the 'Remarks on the Legend of the Giants Following Hebrew Etymology,' in which he analyzed the etymology of names from Greek myths, tracing them back, often wrongly, to origins within the Oriental languages he was studying. If the result was flawed, the methods fascinated him, and he decided the best way to study Antiquity was through language. Even though he had been studying Egypt for a few years, hieroglyphs had not yet begun to obsess Champollion – his real passion was still the chronology and origins of mankind. To this end he read voraciously, completely out-distancing his fellow students, but he began to realize that he might gain access to as yet undeciphered and unknown texts containing the earliest records of human history (earlier even than the Bible) through Coptic (the language used by Christian Egyptians) and through hieroglyphs – a script that had baffled scholars for over a thousand years must be made to give up its secrets.

During the fifteenth and sixteenth centuries a number of ancient Greek and Latin texts had been discovered and published (becoming some of the earliest printed books), which revealed to the Renaissance world commentaries on hieroglyphs by Greek and Roman historians. None of these ancient authors had understood hieroglyphs, but they spread the false notion that the writing system (which largely comprises recognizable pictures of natural or man-made objects) contained symbolic or allegorical messages. In 1419 a Greek manuscript of the *Hieroglyphica* written in the fourth or fifth century by an author called Horapollo was brought to Italy. Its 189 sections each dealt with a particular hieroglyph, and although the author had some genuine knowledge of hieroglyphic writing, he also gave fanciful allegorical interpretations for many of the signs and even invented spurious hieroglyphs. Numerous copies of the manuscript were made and circulated in Florence, and it was first printed and published in 1505, with numerous editions in several languages over the next century. The *Hieroglyphica* led to a fascination for hieroglyphs among artists and intellectuals in Renaissance Italy and beyond. In an age when literally everything – dreams, the landscape, comets – was analyzed for symbols that could be interpreted, hieroglyphs were seen as the key to true knowledge: ancient Egyptian religion was believed to contain prophecies about Christianity and hieroglyphs to be symbols expressing the sacred truths that could not be revealed in mere words, but had to be hidden from the uninitiated.

Over the next three centuries, the idea that hieroglyphs held symbolic meanings rather than information conveyed by writing in the form of words set in train a series of misguided attempts to decipher hieroglyphs, compounded by the ancient writers who implied that the Egyptians had used both a sacred allegorical script and an ordinary script – for which the Greek historian Herodotus used the terms sacred (*hiera*) and common (*demotika*). Unknown to early scholars, simplified cursive (handwriting) scripts had developed from hieroglyphs to enable faster everyday writing of the Egyptian language. Hieratic was the earliest such cursive script and was used during much of Egypt's history. As the Egyptian language changed over time, so did the hieratic script, until by about 650 B.C. both the language and script had changed so much that they are today given the name demotic. A script that is difficult to read and barely recognizable as a descendant of hieratic and hieroglyphs, demotic was also used for some monumental inscriptions, as on the Rosetta Stone, even though hieroglyphs continued to be employed. From the Roman period, a new script known as Coptic was used to write the slowly evolving Egyptian language and comprised a mixture of Greek and demotic letters of the alphabet. Despite the introduction of the Arabic language and writing system into Egypt in the seventh century, Christians continued to speak and write the Egyptian (Coptic) language – 'Copt' simply means 'Egyptian.' It was this Coptic language that was beginning to interest Champollion. In summary, the languages are:

| Written script | Spoken language |
|---|---|
| *Hieroglyphs*<br>(formal writing)<br><br>*Hieratic*<br>(cursive 'handwriting') | Ancient Egyptian |

| *Demotic*<br>(from 650 B.C., used only for the demotic language) | Demotic<br>(a late development of ancient Egyptian) |
|---|---|

| *Coptic*<br>(from A.D. 250, used only for the Coptic language) | Coptic<br>(developed from demotic) |
|---|---|

Hieroglyphs can be thought of as the equivalent of formal printed text, as would be found today in books and on monuments, whereas hieratic script was the equivalent of fine copperplate handwriting. If hieratic is copperplate, demotic is ordinary handwriting: compared with the clarity and beauty of hieroglyphs and hieratic, demotic often appears little better than scribble. Although the Coptic language evolved from demotic, the written form of Coptic is completely different from hieroglyphic, hieratic and demotic scripts, consisting of the letters of the Greek alphabet and a handful of demotic characters. For the first time, vowels were written down.

Following the first publication of the *Hieroglyphica* in 1505, numerous other works on hieroglyphs were produced in Italy, although all accepted the conventional, misguided views on their meaning. Pierio Valeriano became obsessed with the study of hieroglyphs and wrote a huge commentary in fifty-eight books, which he also called *Hieroglyphica*. Published after Valeriano's death in 1558, it divided hieroglyphs into distinctive categories such as pagan gods, parts of the body and plants, and discussed what he considered to be the religious and philosophical significance of each hieroglyph. Despite being no more accurate than any of the preceding works, it was to remain the unchallenged authority on the subject for nearly two centuries, with numerous editions and translations into Italian, German and French. There was no end to the enthusiasm for hieroglyphic symbolical studies, but few works attempted to distinguish between recently invented hieroglyphs and authentic Egyptian ones.

It was not until the seventeenth century that the first real attempts to decipher Egyptian hieroglyphs began, led by Athanasius Kircher, an accomplished Orientalist who had fled to Rome from Germany during the Thirty Years War. A Coptic scholar, he became involved with publishing work on Coptic vocabularies and grammars that had been recently brought back as manuscripts from Egypt. Kircher correctly deduced that Coptic, which was then still used as a liturgical language by the Christian Egyptians, could be the same language as that of ancient pharaonic Egypt and that a knowledge of Coptic was therefore crucial for the understanding and decipherment of hieroglyphs. By using the genuine Egyptian monuments and artifacts that had been discovered or were still upstanding in Rome, Kircher was the first scholar to study genuine hieroglyphs, but continued to pursue the obscure symbolism that hieroglyphs were supposed to contain instead of regarding them as a script, and so his translations are meaningless. One important result of his work was that Coptic studies became fashionable amongst scholars.

The philosophy of Kircher and others, which saw Egypt as the source of all wisdom, was gradually discredited, and travellers began to dispel the notion of Egypt as a mystical land. In 1741 William Warburton published *The Divine Legation of Moses*, which contained a lengthy digression on hieroglyphs that was translated into French three years later. Although he failed to study actual hieroglyphs, some of what this English bishop had to say on methods of decipherment was very near the truth, and had his views been tested the process of decipherment might have been a great deal shorter. He even deduced that what is now called hieratic owed its origins to hieroglyphs. Contrary to Renaissance thinkers, he made the astonishing and yet accurate pronouncement that the script was not a sacred invention of the ancient Egyptians 'in order to hide and secrete their wisdom from the knowledge of the vulgar.' With this vehement attack on Kircher and his mystic beliefs, Warburton heralded a new era of scientific ideas on hieroglyphs.

Some twenty years later, Abbé Jean-Jacques Barthélemy, Keeper of Medals at Paris, was the first to state that the oval ring with a bar at one end (nowadays called a cartouche), often found within hieroglyphic inscriptions, could contain royal names or those of deities. A few years afterwards this idea was developed by Joseph de Guignes, Professor of Syriac at the College of France in Paris. Through his Chinese studies, de Guignes recognized that cartouches in Chinese writing were used to highlight proper names and that they were therefore probably used for royal names in Egyptian inscriptions. Unfortunately this coincidence led him to become bogged down in the strange idea that China had been an Egyptian colony and that while the Egyptian language had itself later become corrupted by Greek, the Chinese language and writing represented the true uncorrupted form of Egyptian. He therefore maintained that Chinese and not Coptic was the way forward in hieroglyphic decipherment, an idea that gained enormous popularity and sent many potential decipherers heading in totally the wrong direction. The publications on hieroglyphs available to Champollion covered a motley range of theories since a possible connection between Egyptian hieroglyphs and Chinese writing had still not been disproved.

Of more recent work, the most important was that of the Danish scholar and expert in Coptic, Georg Zoëga. In 1783 he had gone to live in Rome and undertook a major study of its obelisks, making important observations on hieroglyphs and criticizing many of the misconceptions of his predecessors. Zoëga was the first to compile a corpus of hieroglyphs, recognizing 958 in all, which he grouped according to what they depicted, such as plants, utensils and parts of mammals, rather than their meanings, which were still completely unknown. For example, 𓀀𓀁𓀂𓀃𓀄 depict different aspects of people and 𓅿𓆀𓅪𓅬𓅭𓅮 depict various birds. He also made the important observation that the direction of reading an inscription depended on the direction that the signs faced, so that the hieroglyphic phrase 𓋹𓍑𓋴𓄿𓈖𓌅𓏴 which means 'may he live, be prosperous, be healthy' is read from left to right, since the hieroglyphs are facing left. The same phrase can be written as 𓏴𓌅𓈖𓄿𓋴𓍑𓋹 which has exactly the same meaning, but is read from right to left because the hieroglyphs are facing right. Hieroglyphs always 'look' towards the beginning of a line of text, and vertical columns of hieroglyphs are always read from the top to the bottom. Zoëga did not undertake any decipherment of hieroglyphs himself, but repeated the belief that cartouches contained proper names or religious formulas, a statement that would finally lead other scholars to investigate this aspect of hieroglyphs in relation to the Rosetta Stone.

Attempts at deciphering hieroglyphs had remained in limbo until the work of the savants participating in Napoleon's Egypt expedition became known, boosted by the discovery in 1799 of the Rosetta Stone. It was only as Napoleon's army fought its way into the heart of Egypt, with the savants trailing behind, that they were to discover the huge variety of hieroglyphic symbols, most of which had never been seen outside Egypt, and scholars realized that the previous attempts to decipher hieroglyphs from inscriptions on a tiny number of monuments in Europe had been futile. The Rosetta Stone is bilingual, having three inscribed texts but just two languages – the Greek was written in upper-case letters, the Egyptian was written in hieroglyphs (regrettably the most damaged part of the stone), and the third inscription was a later form of Egyptian written in the demotic script. Copies and casts of the Rosetta Stone brought back by the savants from Egypt had already begun to be studied in Paris, and the Greek text had been translated. The contents of the texts on the Rosetta Stone are not in themselves important, or indeed interesting, being a record of a priestly decree in 196 B.C. setting up a cult of the reigning king, the Macedonian Greek Ptolemy V Epiphanes. Mainly consisting of a lengthy hymn of praise, it begins: 'In the reign of the young one who has inherited the kingship from his father, Lord of Crowns, great of glory, who has established Egypt and is pious towards the gods, triumphant over his enemies, who has restored the civilized life of mankind, Lord of the Thirty Years Festivals ...' and continues in a similar vein. The importance of the texts lay in the fact that the same message appeared to have been written in hieroglyphs, demotic and Greek, and so offered the possibility of clues to deciphering the demotic and hieroglyphic scripts.

In 1802, when Champollion was still only eleven years old and had not even started at the *lycée* at Grenoble, the Oriental scholar Silvestre de Sacy in Paris decided to tackle the Rosetta Stone. He began by trying to identify in the demotic script the proper names that occurred in the Greek text, but was only barely successful in locating the approximate groups forming the names 'Ptolemy' and 'Alexander.' Admitting defeat he passed his copy of the Rosetta Stone texts to his student Johan David Åkerblad, formerly a Swedish diplomat at Constantinople whose main interest was in languages. Incredibly, within two months Åkerblad succeeded in identifying all the proper names within the demotic text that were also present in the Greek text, and showed that they were written with phonetic alphabetic symbols – signs that represent a single sound, like letters of the alphabet. Names such as Ptolemy, Cleopatra, Alexander, Berenice, Arsinoë and Alexandria could now be read in demotic. By applying his knowledge of the Coptic language to the demotic, Åkerblad identified other words, such as 'temple,' 'Egyptian' and 'Greek.' He demonstrated that some words were fairly similar in Coptic and in demotic, which proved that Coptic really was a survival of the ancient Egyptian language, and he established an alphabet of twenty-nine demotic characters, although half turned out to be wrong. His results were presented in the form of an extended letter to Silvestre de Sacy (*Lettre sur l'Inscription Égyptienne de Rosette, Adressée au C.$^{en}$ Silvestre de Sacy*), published in 1802. De Sacy criticized some of Åkerblad's findings in a reply, but ended with the encouraging words, 'If you are determined to publish the Letter which you have done the honour of addressing to me, I would be flattered if you would join my reply to it; that would assure me the advantage of having praised your work the first.'

Åkerblad and de Sacy failed to make further advances within the demotic text because they were convinced that the entire demotic script was alphabetic, just like the Greek script. Apart from some minor work on the numerals, they did not even reach the stage of studying the hieroglyphs. Still no one could read a word of hieroglyphs, but the Swedish Orientalist scholar and diplomat Count Nils Gustaf Palin did attempt to decipher those on the Rosetta Stone, albeit working from a poor copy of the inscription. He published several papers on the subject around the time that Åkerblad was working, but his theories were as fantastic as de Guignes' earlier idea that China had been a colony of Egypt. Claiming that Chinese and Egyptian hieroglyphs were identical in origin and meaning, Palin said that 'we have only to translate the Psalms of David into Chinese, and to write them in the ancient character of that language, in order to reproduce the Egyptian papyri, that are found with the mummies.' Tragedy struck him in later life when he was murdered in Rome while trying to defend his rich collection of Egyptian antiquities from robbers.

During the early years of the nineteenth century no real progress was made on the decipherment of hieroglyphs, although more and more people attempted to tackle the problem, which was turning into an unofficial race with no starting line and with competitors very often unaware of each other. Some, like de Sacy and Åkerblad, even dropped out before the race really started. Hieroglyphs were a popular topic and 'much was written and said about

them by the faddists and cranks, who were usually wholly unedu-
cated men.' The early promise of the Rosetta Stone proved to be a
disappointment, and the initially intense interest in it began to
wane. When the stone was discovered, the savants had assumed
that only a few weeks' investigation would be needed to force the
stone to give up its secrets, but this had not happened. A swift reso-
lution was unlikely, but this did not prevent scholars continuing to
discuss the stone as if it was the only means of deciphering hieroglyphs. Even Jacques-Joseph was to study it and presented a paper
on its Greek inscription to the Academy of Grenoble in the summer
of 1804.

While working in his cousins' firm in the Grande Rue, Jacques-
Joseph had continued his own self-education as best he could. He
collected even more books and frequented the educated and literary
circles within the Grenoble bourgeoisie. By now an accomplished
linguist, he was also deeply interested in ancient history in general,
and during their years at Grenoble the concerns of the two brothers
ran roughly in parallel. From his studies of ancient buildings in the
town, Jacques-Joseph had gained a reputation as a skilful antiquarian and had come to the notice of Joseph Fourier who, being a
veteran savant from Napoleon's Egypt expedition, maintained a
keen interest in many aspects of history. As Prefect of the Isère, Fourier put Jacques-Joseph in charge of recording ancient Latin
inscriptions when they came to light during building work in Grenoble, which was on top of the site of a Roman town called
Gratianopolis, and soon afterwards, at the end of 1803, Jacques-
Joseph was accepted into the Académie delphinale. Like all learned
societies in France, the Academy was suppressed by decree during
the Revolution, but had been resurrected in 1795 under the title
'Society of Arts and Sciences of Grenoble,' later reverting to its original name. The Academy was the most prestigious learned society
in the region, respected throughout France and abroad, and the
acceptance of Jacques-Joseph Champollion-Figeac, who was only
twenty-five years old, who did not have a university education and

had only been in Grenoble for five years, was a measure of his abilities and enthusiasm. Through membership of the Academy and friendship with Fourier, Jacques-Joseph furthered both his academic career and his social standing.

Due to his association with Fourier, Jacques-Joseph also became deeply involved with the work being carried out to prepare the Egypt expedition's *Description de l'Égypte* for publication by helping to research parts of the extensive preface that Fourier was writing. Champollion was also involved in this research, preparing reports on various subjects, but it appears that Fourier was at first unaware of this. By 1804 both brothers were immersed in the study of ancient Egypt, and in that year Jacques-Joseph was made joint secretary of the Academy. Two years later he was made full Secretary, a post to which he was repeatedly re-elected for the next ten years, and he published a paper about Egypt called '*Lettre sur une inscription grecque du temple de Denderah*' (Letter about a Greek Inscription from the Temple at Dendera) which was formally addressed to Fourier.

Champollion himself was becoming increasingly absorbed by all aspects of ancient Egypt, but from his reading of the work of previous researchers, he had become convinced that Coptic was definitely related to the ancient Egyptian language, and so his immediate goal was to learn Coptic – no easy task within the confines of the *lycée*. In 1805 he met a former Coptic Christian monk called Dom Raphaël de Monachis, who was visiting Grenoble. Having returned with the Egypt expedition, he had been appointed by Napoleon as a lecturer in Arabic at the School of Oriental Languages in Paris. During his stay in Grenoble he helped Champollion to learn Coptic and returned months later bringing a number of books, including a Coptic grammar. By now Champollion seems to have settled on Egypt as his course of lifetime study, fervently declaring his allegiance to that country: 'I wish to make of this ancient nation a profound and continuous study...Of all the peoples that I love best, I swear to you that nobody displaces the Egyptians from my heart!'

At this time he was barely aware that rivals might be working hard to decipher hieroglyphs, and in any case the thought of rivals in his quest was insignificant compared to his relationships with the teachers and his colleagues. He had at least one close friend at the school, called Johannis Wangehis, but we might not know even his name if one of the teachers who disliked Champollion had not decided to keep the two apart as much as possible. Pouring out his resentment in his letters to his brother, he wrote bitterly of what he perceived to be the petty vindictiveness of his teachers:

My friend took no account of their advice. He was always my consolation, and these monsters irritated at seeing him still with me have just changed his course of study. I will no longer see him except in passing...My head is no longer my own – I am furious. When will my torment end?...If anybody is frustrated or unhappy in this *lycée*, it is *me*! They will make me lose my head.

Some later commentators have suggested that there was a homosexual relationship between the two adolescents, particularly since in an angry letter to his brother Champollion complained that the teachers had warned Wangehis not to keep company with him any longer because he was being 'corrupted.' It is perhaps more likely that some of the less able teachers found Champollion too talented and troublesome to deal with and wanted to curb his influence as much as possible. In his letters to Jacques-Joseph, the separation from Wangehis is mixed up with other complaints about the staff, and Champollion denounces a teacher whom he regards as a bigot and a hypocrite who even dares to speak ill of Jacques-Joseph. So much a feature of his later life, Champollion's flair for making as many bitter enemies as he made faithful friends had already evolved – his lack of tact and his directness in giving both praise and criticism sometimes caused offence where none was intended.

Some months after the two friends were separated, Wangehis fell ill and eventually had to leave the *lycée* to be cared for by his family. Champollion was also not fully fit, despite school reports which gave him excellent marks for health. Apart from problems with his eyesight due to excessive illicit reading in dim light and exhaustion from too little sleep, his letters to Jacques-Joseph often contain complaints of feeling ill and descriptions of various symptoms. It is difficult to estimate the severity of these illnesses, but since his complaints were often coupled with pleas to be released from the *lycée*, at least some of his ailments seem to have been exaggerated. Other letters to Jacques-Joseph are full of apologies for his attitude and for not working harder, and of gratitude for all that his brother is doing for him. During his entire time at the *lycée* he was in an emotional turmoil. He knew his brother had made many sacrifices to get him into the best available school and to help with his private studies, and this put pressure on him to make the most of the opportunity, yet he hated the discipline, the fanatical adulation of Napoleon that pervaded the *lycée*, the pettiness and spite of some of the staff, the suffocating restrictions and above all the lack of freedom to study his own chosen subjects.

In August 1806 Champollion was chosen to make a public speech at the end of the school year in front of the Prefect, Joseph Fourier, in order to demonstrate the high quality of education provided by the *lycée*. Petrified at the prospect of speaking in front of so many people, he wrote to his brother begging him to do something to prevent it: 'I am really sorry about the honour which Monsieur the Prefect wishes to give me, but I believe it is impossible that I can conquer my timidity. If I am troubled in front of four people, how much more reason in front of a thousand. I beg you to do everything possible so that it doesn't take place.' In the event he overcame his timidity, his explanation of a part of Genesis in Hebrew was very successful, and it was reported in the local newspaper that the Prefect had shown great satisfaction in his performance – despite his diffidence, Champollion's unusual talents were beginning to be widely recognized. Nearly sixteen when

he returned to the *lycée* in November, he had by then made remarkable progress both with his linguistic studies and his research on
Egypt. Although still emotional in tone, his pleas to his brother to
be allowed to leave the *lycée* grew more insistent and more rational,
and Jacques-Joseph began to take steps to enable Champollion to
study at Paris. His escape could not come too soon, but he was bitterly disappointed to learn that he would have to endure another
year at the *lycée* before he could transfer to the capital.

The distant prospect of being able to leave the *lycée* for Paris
seems to have given him new strength, and he plunged into compiling a *Geographical Dictionary of the Orient* from the material he had
been accumulating over the years – a dictionary being an appropriate type of book for Champollion to write considering his enthusiasm for collating and classifying data. In order to read everything
he could about Egypt, in whatever language it was written, he
began to teach himself Italian, English and German, the latter being
the one language that he could not master. The research for his dictionary extended to the Bible, which he read not with the eye of
faith, but with the critical analysis of a historian. He doubted that
Moses, brought up in Egypt, was the author of the first five books of
the Bible as was traditionally supposed, because they were not written in his own native Egyptian. At the same time Champollion
began to compile whole bundles of notes on 'the symbolic signs of
the Egyptians,' studying in particular the *Hieroglyphica* of Horapollo, which although highly unreliable, was still the best available
ancient text on the meaning of hieroglyphs.

This intensive research was disrupted early in 1807 by a revolt
of the boarders at the *lycée*. Apparently sparked by a harsh and
unjust punishment of some of the pupils, a group of the boarders
gathered sticks and stones during the day and rioted at night. Ston-
ing the dormitory windows, they smashed all the glass and then
hammered the casements with their chamber-pots. Appeals by the
assistant head of the *lycée* failed to calm the situation, and the riot
was only stopped by troops with fixed bayonets, who were then set

to guard the dormitories and keep order. In a letter to his brother, Champollion protested his innocence ('I didn't take any part in it'), but given his hatred of the *lycée* and the fact that the rest of the letter is an eyewitness account of the uprising, this is difficult to believe. Whatever the truth, he does seem to have benefited by the riot, as he once again begged his brother to remove him from the *lycée*, and this time he was supported by the Prefect Fourier. Jacques-Joseph took him out of boarding, and Champollion once more lived with his brother, only attending selected classes at the *lycée*. He now had more time and more freedom to follow his own studies, and his progress accelerated.

It was arranged by Jacques-Joseph that Champollion would go to Paris to study in the autumn of 1807, but his long summer of intensive preparations was interrupted by a succession of events. On 19 June, just after he had written to ask his parents to come and see him before he departed for Paris, his mother died. In the following month the two brothers spent some days in Provence and met up with their widowed father at the fair at Beaucaire, which he attended every year to buy and sell books. From early in the sixteenth century such fairs had been essential for travelling salesmen from all over western Europe who began to concentrate their activities at the Frankfurt book fair, which became the focus of the international as well as the German book trade. By about 1750 the prestige of Frankfurt had declined and Leipzig was taking its place – virtually no French publishers were now represented at Frankfurt, but concentrated their business in Paris and regional fairs such as that at Beaucaire.

When the brothers met their father, Jacques-Joseph was newly married, the wedding having taken place at the beginning of July, to Zoé Berriat, daughter of an established bourgeois family in Grenoble. Her father was president of the council of advocates in the town and although the family was not particularly wealthy, her dowry was a country house at Vif, just south of Grenoble. Despite his close relationship with his brother, Champollion was perhaps surprisingly not jealous, but approved of the marriage and had a good relationship with his sister-in-law. She was able to tease him without provoking his quick temper, and she flattered him by protesting against him being called *cadet* (junior) because he was the youngest in the family. He had taken to signing *cadet* instead of his first name, but having learned that the Arabic *seghir* had almost the same meaning, Zoé insisted that he change his name, and from then on he was always known to family and friends as *Seghir*.

The 27th August 1807 was a day of celebration for pupils at the *lycée*, marking the end of the school year. Champollion was awarded a 'certificate of study and conduct' and had already been accepted as a student in Paris. His own cause for celebration was not his past achievements, but his dreams of the future and above all his final release from the place he hated so much. He was so overwhelmed with emotion during these celebrations that he collapsed in a dead faint. Just as his lessons at the *lycée* had come to take second place to his own research, the school certificate meant little to him compared to the recognition he was to receive five days later from the Académie delphinale. In front of an audience representing the elite of the scholars and intellectuals in the region, the sixteen-year-old presented his 'Essay on the Geographical Description of Egypt before the Conquest of Cambyses.' His extraordinary gift for languages was by this time well known, but this essay was the first real evidence of his ability to carry out original research, collate and analyze the evidence, and present the results in an intelligible way. Those present were astounded at the quality of the work and immediately proposed that Champollion should be admitted as a member of the Academy – a remarkable honour that was formally ratified six months later and with which he was overjoyed. He wrote to Jacques-Joseph: 'My membership of the Academy of Grenoble gives me a keen pleasure. What delights me the most in that, is that I am *a little more your brother*.'

# CHAPTER THREE

# (The City)

Champollion and Jacques-Joseph set out from Grenoble early on 10 September to travel over 300 miles to Paris in one of the heavy horse-drawn public carriages known as diligences – Champollion would not see Grenoble again for over two years. After a wearying continuous journey by night and day across a large tract of France, on dreadful roads, with little sleep and the frequent danger of robbery, they arrived in Paris on the 13th. On the day after their arrival, the future of Egypt was settled when the British signed a treaty of evacuation. Following the repatriation of the French Egypt expedition towards the end of 1801, a struggle for power in Egypt had involved the Turks and the British as well as several Egyptian factions. Having left Egypt soon after the defeat of the French, the British had been persuaded to reinvade to support the faction of the Mamelukes – the invasion was a fiasco. Eventually negotiations between the British, the Mamelukes and a force of Albanian mercenaries commanded by Mehemet Ali led to the British evacuation and gave Mehemet Ali a tenuous control of Egypt on behalf of the Turks.

Once Champollion arrived in Paris, having spent so long hoping to escape from the *lycée*, perhaps it was inevitable that his high expectations of the capital would be disappointed. It was not his teachers or the facilities for study that failed to please him, but the city itself: after the grandeur and beauty of the mountains around Grenoble and the picturesque splendour of the valley of the River

Célé at Figeac, he found Paris noisy, dirty and squalid. Also, moving from the claustrophobic community of the *lycée* to a city where he was yet to make friends, he was touched by homesickness, poignantly writing to Jacques-Joseph, who had returned to Grenoble: 'I have never left you, and now I am alone; write to me often.' More seriously, the move from the clean air of the mountains to the swampy lowland vapours of Paris immediately affected his health and his state of mind: 'I have had quite a pain in my side. The air of Paris erodes me, I spit like a madman and I am losing my strength. The country here is horrible, one always has wet feet. Rivers of mud (without exaggeration) run in the streets, and I am completely fed up with it.' Paris at this time was a ruinous, sordid place, with some buildings in disrepair, others empty and stripped of materials. Historic monuments and buildings were vandalized or demolished, and streets were extremely narrow, ill-lit, filthy and evil smelling, with a lack of pavements, drains and sewers. Every kind of detritus imaginable would pile up in the streets, much washing into the Seine, a river with few embankments and few bridges that acted as a huge open sewer. Paris would not be transformed into a city of wide elegant boulevards until Josephine's grandson ruled France decades later – the Paris Champollion knew was dank, dirty and disease-ridden, with much of the damage caused during the Revolution still awaiting repair.

By contrast, intellectual and cultural Paris was without parallel. At the *lycée* Champollion had been irritated by the fanatical homage to Napoleon and already held anti-imperial opinions, yet this imperialism which he so despised had opened Egypt to the western world and made Paris the centre of scholarship in Europe. The libraries in Paris were bulging with books and manuscripts looted from the territories Napoleon had captured, while public museums and private collections were stuffed with the art treasures of Europe – Champollion informed his brother that November: 'There has lately been an exhibition at the Napoleon Museum of a beautiful collection of wonderful pictures, conquered in Germany, Prussia

and Russia, as well as fragments of many pieces of antiquities and
Egyptian statues.' Napoleon placed particular importance on the
Institute of France, a prestigious body where the latest research was
constantly presented by its eminent scientists, who were regarded
as public servants and paid a modest stipend. Paris was already an
important centre for many academic disciplines, but now it
attracted the best and most famous scholars and teachers in
France, as well as many from other countries, making it the vibrant
focus of all scientific and artistic innovation and the heart of fashionable Europe. Even the English, in the short cessations of
hostilities that punctuated the war with France, were drawn to the
city in droves, and in 1814, when Napoleon was first forced into
exile, so many visitors were present that there was even a popular
song, 'All the world's in Paris.' In the years before Napoleon was
defeated and much of the loot was repatriated across Europe, in particular to Italy and Germany, Paris offered the best facilities for
study and research: Champollion was fortunate to be able to study
there at the peak of its prestige.

His dislike of Paris at least had the advantage that he was seldom
diverted by other attractions but immersed himself in study. He
divided his time between the College of France, the Special School of
Oriental Languages, the National Library and the Commission of
Egypt responsible for the publication of the *Description de l'Égypte*.
Within a few months he had built up an exhilarating and exhausting regime, travelling back and forth across the Seine – a routine
which he outlined in a letter to his brother:

On Mondays, at 8.15, I leave for the College of France where I
arrive at 9am. You know what a long way that is – it is in the
Cambrai Square, near the Pantheon. From 9am to 10am I
follow the Persian course of Monsieur de Sacy. Leaving the
Persian course, as the Hebrew, Syriac and Chaldean course
starts at midday, I go immediately to Monsieur Audran, who
has proposed reserving Mondays, Wednesdays and Fridays,

from 10am to midday for me. He resides inside the College of France. We spend these two hours talking about Oriental languages, and in translating Hebrew, Syriac, Chaldean or Arabic. We always devote half an hour to working on his 'Chaldean and Syriac grammar.' At midday we leave and he teaches the Hebrew course. He calls me the 'patriarch of the class' because I am the most able. After this course, at 1pm, I go right across Paris to the Special School for the start of the course of Monsieur Langlès at 2pm, who gives me particular attention...On Tuesdays at 1pm, I go to Monsieur de Sacy's course at the Special School. On Wednesdays I go to the College of France at 9am; at 10am I go up to see Monsieur Audran. At midday, I go to his course. At 1pm I go to the Special School for the course (two hours) of Monsieur Langlès; and in the evening at 5pm I follow that of Dom Raphaël, who makes us translate the fables of La Fontaine into Arabic. On Thursdays at 1pm the course of Monsieur de Sacy. On Fridays I go as on Mondays to the College of France and with Messieurs Audran and de Sacy. On Saturdays, with Monsieur Langlès at 2pm.

His remaining time was spent on his own research, visits to the church of Saint-Roch to improve his Coptic with the priest there who spoke that language and running various errands for his brother.

At long last Champollion felt free to attend the courses that really inspired him and to pursue his own research without being forced to study subjects that bored him. He began to show the ability to focus on the essential elements of a project and the tenacity of purpose that would later enable him to continue his research despite many setbacks and interruptions. His studies at Paris not only provided him with the knowledge and skills he needed, but also contacts and often friendships with many of the most eminent linguists and Orientalists. At that time two of his professors, Louis

Langlès and Silvestre de Sacy, were the foremost Oriental scholars in Europe. Antoine Isaac Silvestre de Sacy, despite being a Royalist and a supporter of the Catholic Church, had managed to survive the Revolution and had been made a Knight of the Legion of Honour by Napoleon in 1803. Three years later, at the age of forty-seven, he was appointed to the chair of Arabic at the College of France. An extremely gifted linguist, he taught Champollion Persian and Arabic, but was also a major influence on him and encouraged him to pursue his own research. Although de Sacy's attempts to study the Rosetta Stone had met with minimal success, he maintained a keen interest in hieroglyphs and was at the centre of all scholarly discussions and controversies about ancient Egypt. At first overawed by him and timid in his presence, Champollion quickly established a rapport with de Sacy, regarding him as more modest than Langlès but also as somebody who never made friendships with anyone. For his part, de Sacy would later record that 'my personal memories recall still the first meeting, which left profound impressions in my mind. Needless to say the new pupil was faithful to his vocation, and followed assiduously the lessons which he had come to seek in the capital.'

Champollion's relationship with Langlès was not so cordial. Louis-Mathieu Langlès was one of the founders of the Special School of Oriental Languages, where he taught Champollion Persian. Himself once the pupil of de Sacy, he was disdainful of everybody and only favoured those students who committed themselves wholeheartedly to studying Asian languages. At first Langlès tried to deflect Champollion's interest from Egypt to Asia, but having failed in this, Langlès became increasingly hostile towards him, and the hostility was rapidly returned. With his instinct for barbed and witty nicknames, Champollion referred to Langlès as 'l'Anglais' ('the Englishman') – a particularly pointed insult in view of the continuing war between France and England.

The great Hebrew scholar Prosper Audran taught Champollion Hebrew, as well as related languages such as Aramaic, and was very impressed with his exceptional command of languages. A deep friendship developed between the two, and Audran gave Champollion extra tuition and let him help with a book of Syriac grammar and a comparative grammar of Arabic and Hebrew that he was compiling. It is a mark of Audran's complete confidence in Champollion's ability that he sometimes set the seventeen-year-old to teach Hebrew to his fellow students, who were largely members of the clergy learning the language in order to study the Bible. Few of them had a gift for languages and Audran's nickname for Champollion – the patriarch – must have been particularly galling to them. In high spirits, Champollion wrote to his brother that 'I have already gained the favour of M. Audran our professor: "you are young, you have courage, we will be able to do something useful." These are his words...he shows me a lot of friendship.'

Having helped Champollion study Coptic in Grenoble, Dom Raphaël de Monachis now taught him Coptic and Arabic, with Champollion self-mockingly recording that 'I already *cough up* Arabic very prettily.' By the end of the year his intensive study of Arabic had affected him to such a degree that he noted, 'what is bad is that Arabic has totally changed my voice, has made it muffled and guttural; I speak without scarcely moving my lips.' Dom Raphaël was also responsible for introducing him to Geha Cheftitchi, the priest of the church of Saint-Roch, with whom he practised speaking Coptic and obtained much information on the Coptic names of people and places in Egypt. By now Champollion was well aware that others were working on hieroglyphs, but with the confidence of youth he was convinced he would succeed, and for the time being he decided to concentrate on Coptic and other Oriental languages, for which he had enormous talent and enthusiasm: 'M. Langlès is rather pleased with me regarding Persian, which I translate very easily. If Arabic is the most beautiful of languages, then Persian is the sweetest; I have sweated blood and tears in sorting out Ethiopian, and I have succeeded. I have studied its links with Hebrew and Arabic, and I am in a position to translate it with much ease.'

While Champollion was a student in Paris, the learning of Oriental languages in order to study earlier and seemingly more accurate versions of the Bible was not regarded as an obscure and esoteric occupation, but the cutting edge of research. The separation of various branches of scholarship was only just beginning, and the great split between sciences and arts was yet to take place. The date of the creation of the world and its early history still relied totally on the chronology deduced from the Old Testament books of the Bible, and this chronology would remain almost unchallenged until some of the scholars studying Egyptian antiquities began to suspect that they dated to a time before that supposedly recorded in the Bible – before the creation of the world was supposed to have taken place, which was a potentially shocking concept.

Through his contact with the Prefect, Joseph Fourier, in Grenoble, who was still working on part of the *Description de l'Égypte*, Champollion also became involved with those working on it in Paris, and was introduced to Edme-François Jomard, the engineer, geographer and antiquarian who had taken part in Napoleon's Egypt expedition and had assumed responsibility for editing the *Description de l'Égypte* after the death of the first two editors; the initial volume was yet to be published even though work on it had been continuing for some years. When they met, Champollion made no secret of his ambitions and presented Jomard with a copy of a map of ancient Egypt that he had compiled as part of his presentation to the Academy at Grenoble. Jomard, who was himself working on the geography of Egypt and had his own aspirations to decipher the hieroglyphs, was deeply offended by the young student's presumption – he took an instant and irrational dislike to Champollion, recognizing him as a dangerously talented rival. Despite remaining on friendly terms with his brother Jacques-Joseph, Jomard was to prove a lifelong enemy of Champollion, obstructing his progress whenever he could. Champollion was on much better terms with Prosper Jollois and Édouard de Villiers du Terrage, the two engineers who had played a large part in recording sites and monuments during the Egypt expedition. These contacts with the scholars working to produce the *Description de l'Égypte* kept him informed of the latest theories on Egypt, as well as providing, well before publication, first-hand access to the drawings done by the savants and engineers of the hieroglyphs carved on the temples and tombs – an advantage over most of his rivals.

Using every waking minute of the day for study or research, and often having to run through the streets of Paris from one lecture to another, it was not long before Champollion's health began to suffer. He had violent headaches, pains in various parts of his body, difficulty breathing and a cough. Feeling tired and overheated were frequent complaints, for which he was prescribed sweet cooling drinks that had some beneficial effect. The reports about his health in letters to his brother at this time were in a different vein to those in his letters from the *lycée*: the letters from Paris are more dispassionate, and although eliciting sympathy they are no longer coupled with pleas to remove him from a difficult situation. Nevertheless, he became so thin by the end of 1807 that those around him began to suspect he had tuberculosis, and in July the following year he declared that his even thinner appearance and hollow cheeks made him look still more like an Arab.

His ill-health does not seem to have affected his studies seriously, probably being more of a nuisance than a handicap, and possibly caused largely by his poverty. His stay in Paris was even more impoverished than his time at the *lycée*, because although Jacques-Joseph had applied for an extra government grant and tried to find Champollion a job in the National Library, he had been unsuccessful. The student bursary he received from Grenoble still only covered about three-quarters of his expenses, and the rest had to be made up by Jacques-Joseph. Champollion certainly did not live in luxury, but he was not always good at making the best of what he had: while his letters to his brother were peppered with constant requests for money, the letters to him from Jacques-Joseph were equally spiced with admonitions to be more careful with his money and with complaints about him spending too much. While Jacques-Joseph was more prosperous than he had been when his brother was at the *lycée*, the cost of supporting Champollion in Paris was nevertheless a difficult burden.

Jacques-Joseph now had a son, but before he was born Champollion was asked to choose an Arabic name: 'It did not take me long to seek the Arabic name of my future nephew: he will be called Ali (the beloved), a name which will give no offence to French ears. If it is a niece, her name will be Zoraïde (flower of the spring), which will be the same time when she is to be born.' Years later Champollion would use the name Zoraïde for his own daughter. Jacques-Joseph had even less money to spare, and he seems to have expected his brother to run many errands for him in Paris. By the autumn of 1808 friction was building up between the two brothers. Champollion was becoming desperate: his clothes were little better than rags and he was embarrassed to appear in public. He wrote to Jacques-Joseph: 'The trousers are unwearable; those of nankeen I have used since the summer, and here I am a real *sans-culottes* without however having either the principles or the intentions...When I am clothed and shod I will do all the jobs for you, whatever they may be, for then I will be able to appear in public. I will even go to see the Emperor if that pleases you.' The extreme Revolutionaries, the *sans-culottes* (literally 'without breeches'), had adopted the name because *culottes*, short breeches reaching down to the knee, were worn by aristocrats. The *sans-culottes* were still vividly remembered, but were no longer a dangerous force in France, and people could now joke about them without fear of reprisals. Unable to pay his rent to his landlady Madame Mécran for his room in the rue de l'Échelle-Saint-Honoré, Champollion next sent urgent appeals to Jacques-Joseph for funds: 'Mme Mécran torments me, requesting the payment for the rent of the room' and 'Send money quickly if you have not done so.' Matters came to a head when in despair he was forced to write to his father asking for money. The money was forthcoming, but when Jacques-Joseph found out a bitter argument ensued between the two brothers with recriminations on both sides. Rather than cause a permanent rift between them, the crisis seems to have put their relationship on a firmer footing, and all the problems were settled by the end of the year.

Although Champollion's indifferent health did not prevent him from working intently, and poverty was an irritant to which he was accustomed, the prospect of conscription into the army did frighten him. When Champollion had first arrived in Paris, Napoleon was still at the height of his power, but now this was gradually waning, and he was conscripting increasing numbers of young men to keep his armies up to strength. From the time he reached his seventeenth birthday on 23 December 1807, Champollion was constantly sick with worry about being conscripted. Jacques-Joseph appealed to the Prefect Fourier, who himself approached his friend Antoine Fourcroy, the Director of Education, to intervene on Champollion's behalf, which resulted in the threat of conscription being lifted. By the end of the following summer Napoleon, again short of troops, started another round of conscription. In a panic Champollion wrote to his brother, but Jacques-Joseph's reply was confident, telling him not to worry. Once again he used his influence with Fourier, who applied to Napoleon himself for this student to be exempted 'in the interests of science' – once again Champollion was not drafted.

As well as his regard for the two brothers, Joseph Fourier was by this time greatly indebted to both of them for all the work that they had done on his lengthy preface to the *Description de l'Égypte*. Nevertheless, Jacques-Joseph was aware that the Prefect's influence might not always be effective and advised Champollion to apply for entry to the Normal School, a newly created establishment offering free courses lasting two years, after which the students had to spend at least ten years in teaching. The benefit was that students at the Normal School, unlike other students, were automatically exempt from conscription; the disadvantage was the rigorous military-style discipline with which the school was run. To Champollion the mindless routine of the Normal School was little better than the mindless discipline of the army, and, worse, it would prevent him for many years to come from carrying on his research or doing any of the things he dreamed of, such as travelling to Egypt. He refused to apply, preferring to live under the shadow of conscription and the hardship and danger that might result.

Not only were young men being conscripted into the army, but suitable students from the School of Oriental Languages at Paris were being urged to fill positions as French Consuls. Langlès, who was teaching Champollion Persian and who himself disliked travel (even having refused to participate in Napoleon's Egypt expedition nearly a decade earlier), tried to persuade him to become a Consul in Persia, but Champollion made excuses and refused. From then on he did his best to avoid Langlès, only to find out a few months later in March 1808 that Langlès had put his name forward for the post anyway. As he explained to his brother, the only place where he would want to be Consul was Egypt: 'It is in a far more deplorable state than the fields of Constantinople, Troy and Persepolis: nevertheless, it would offer me enough powerful attractions to make me brave the dangers.' Unfortunately the post of Consul in Egypt was not on offer, and once more Jacques-Joseph had to use his influence to extricate Champollion from a difficult situation. Langlès was very bitter that he did not take the post and out of spite he even refused to give Champollion a certificate of study for that year.

Particularly in the holidays when he stayed in Paris, not having sufficient money to return to Grenoble, Champollion spent much time doing his own research – one major project being the expansion of the work on the geography of Egypt that he had outlined to the Academy in Grenoble in August 1807. By the end of that first year in Paris, he had completed the first draft of the book, which he now called *L'Égypte sous les pharaons* (Egypt under the Pharaohs), based especially on his ever-increasing knowledge of Coptic place names. Much of his research was undertaken in the National Library, where he was greatly assisted by Aubin-Louis Millin de Grandmaison, Keeper of Antiquities, who had been a regular correspondent with Jacques-Joseph for several years and was editor of the journal *Magasin encyclopédique*. At this library Champollion was privileged to gain access to the mass of foreign books that had come as loot from Napoleon's campaigns and which were still not properly catalogued. He set himself to study all the Coptic texts, most of which had been taken from the Vatican Library in Rome. Years later, when these manuscripts were returned to Italy, the English antiquarian Sir William Gell remarked: 'As to the Coptic and Champollion, I think there are few Coptic books in Europe he has not examined: a very learned friend of mine told me there is no book in the Vatican in that language, that has not remarks of Champollion in almost every page, which he made when the MSS. were at Paris.' Champollion immediately found that the existing dictionaries and grammar books of the language were inadequate, so while he pored over the texts he also began the mammoth task of compiling a Coptic grammar and a Coptic dictionary as a prelude to his hieroglyphic studies.

At this stage in his research Champollion assumed that hieroglyphs merely comprised an alphabet used to spell out Egyptian words and that there was little difference between the ancient Egyptian language and Coptic. If so, once he had a sufficient mastery of the Coptic language, all he would need to do was find which alphabetic letters of the Coptic script matched which hieroglyphs in order to read all the ancient Egyptian texts. He was later to realize that this theory was completely wrong and that although Coptic was a late development of ancient Egyptian, hieroglyphs were not a simple alphabet, but the theory provided him with a good basis from which to approach the problem of decipherment.

Other languages that Champollion was studying, such as Arabic and Hebrew, only indicated vowels by symbols (diacritics), yet Coptic did have vowels, written using its Greek-style alphabet. His vast knowledge of Coptic would allow Champollion to work out ancient Egyptian words written in hieroglyphs, which did not show vowels – one aspect that makes hieroglyphs difficult to read because

many of the words are written in a form similar to contracted words
in modern languages. For example, the words 'pkg, grg, gdns' do
not make much sense on their own, but within an advertisement
for rented accommodation they are recognizable as contractions of
'parking, garage, gardens.' Similarly, the name of the most famous
Pharaoh, Tutankhamun, is written 〔⊏⊐⚬⚱♀. The transliteration of
these hieroglyphs is *imntwtᶜnh*. The name is made up of three parts
〔⊏⊐, ⚱ and ♀ meaning 'Amun,' 'image of' and 'living,' and trans-
lates as 'the living image of the god Amun.' There are no proper
vowels in the name *imntwtᶜnh* – the *w* is pronounced as 'u' for convenience, and the phonetic signs *i* and ᶜ represent 'weak
consonants' (sometimes called semi-vowels), which are found in
Semitic languages such as Hebrew. The phonetic symbol *h* is pronounced roughly 'kh.' Just as you have to know that 'gdns'
represents 'gardens' and not 'goodness,' you have to know what
vowels the ancient Egyptians used within each particular word in
order to pronounce their language correctly, but this knowledge
has been lost, although Coptic gives clues to it.

    In order to make the names of the pharaohs at all pronounceable,
vowels are now put between the consonants. Purely for convenience
'e' is often used, which would convert *imntwtᶜnh* into 'Emen-
tutenkh,' but sometimes other vowels are used as well to make a
name sound less monotonous, so changing this particular name to
'Amuntutankh.' Ancient Egyptian puts the names of gods first, out
of respect, but today the elements of a word are rearranged in the
order of sense: 'living image of Amun' is 'Tutankhamun.' The situation is further complicated because the name of the Pharaoh
Tutankhamun contains the name of a god mentioned in the Bible
and in ancient Greek texts as 'Amun' or 'Ammon' (although this does
not mean the Egyptians themselves used that pronunciation). For
this reason, that part of the name is usually spelled today as 'amun,'
'amon' or 'amen.' This happens with the names of many pharaohs,
resulting in a variety of spellings and pronunciations, none of which
can be proved to be the correct ancient Egyptian pronunciation.

Apart from his Coptic studies, Champollion was now ready to tackle the Rosetta Stone with its three scripts: an easy task, he believed. During the summer of 1808 he was able to use a copy of the Rosetta Stone which the elderly Abbé de Tersan had made at the British Museum in London. An antiquarian, Charles Philippe Campion de Tersan was the owner of a large collection of antiquities at his home in the Abbaye-aux-Bois near Paris, where Champollion frequently visited him. Looking at the demotic inscription first of all, Champollion worked out the value of a number of demotic alphabetical letters by comparing this script with the Greek version and then using his knowledge of Coptic. To his delight his findings agreed with those of Johan Åkerblad, who had published his results six years earlier in the form of a letter to Silvestre de Sacy, now Champollion's professor.

At the same time, Champollion began to study examples of papyri with their hieratic writing (the hieroglyphic 'copperplate'), in order to compare them with the demotic inscription on the Rosetta Stone, unaware that demotic and hieratic were two different cursive scripts. He even claimed at the end of August to have proved they were the same and to have deciphered a sentence from one papyrus that Vivant Denon had brought back from Egypt. Only a few weeks later, Champollion wrote to his brother that although he had now read one-and-a-half lines of a papyrus and had compiled an alphabet based on the Rosetta Stone, he was absolutely incapable of progressing: 'I can go no further! Groups [of signs] stop me. I have studied them, thought about them for whole days... and I have understood nothing!' He then went on to say that instead he had become totally absorbed by the Etruscans, a once flourishing pre-Roman civilization in Italy, 'because the Etruscans come from Egypt.' In his collection Abbé Tersan had many Etruscan antiquities, and Champollion was fascinated by them, convincing himself – quite wrongly – that the Etruscans were linked to the Egyptians through the Phoenician people in north Africa, and that the Phoenician alphabet was derived from Egyptian writing, itself alphabetic in nature. He began to study all these related topics in an uncontrolled rush of enthusiasm, as well as many other subjects, such as names of ancient musical instruments, much to the exasperation and anger of Jacques-Joseph, who berated him for having abandoned his study of hieroglyphs: 'You have translated one-and-a-half lines – you have an alphabet and you rest there. I no longer recognize you. Where is your passion for all that relates to Egypt?'

It was a bitter blow when at the end of 1808 Marie-Alexandre Lenoir quite unexpectedly published the first of four volumes entitled *Nouvelles Explications des Hiéroglyphes* (New Explanations of Hieroglyphs), claiming to have deciphered hieroglyphs. Lenoir, twenty-nine years older than Champollion, was much more knowledgeable about Medieval France than ancient Egypt. He had managed to save many ancient French monuments from destruction during the Revolution, sometimes at the risk of his life – work so important that the Society of Antiquaries of London later made him an Honorary Fellow. In Napoleonic France he was a potentially powerful man since as part of his function he was curator to the Empress Josephine, and he helped her to select objects from the Museum of French Monuments in Paris to decorate the Château de Malmaison, her country house just west of Paris. Until he could see a copy of Lenoir's publication, Champollion was worried that he had been beaten in his pursuit of the meaning of hieroglyphs, but on reading it he found that the ideas it contained were totally erroneous, not least because Lenoir was still interpreting hieroglyphs as mystical symbols – the same error that had led many earlier scholars astray. He even went as far as to say that Lenoir's explanation of hieroglyphs should serve Jacques-Joseph as a medicine – to cure him of worrying about Champollion's apparent lack of work, or as a purgative.

The experience with Lenoir frightened Champollion out of his complacent belief that he would be the first to decipher hieroglyphs. He suddenly realized that although he was laying solid foundations for his work, he could no longer ignore the fact that he was part of a race against any number of scholars, many not known to him, who might publish their results and claim the prize at any time – rivals who would turn out to be eccentric, obsessed, jealous and even vindictive when Champollion finally succeeded. Nevertheless, he refused to be deflected from his desire to be thoroughly fluent in Coptic, declaring to his brother in March 1809: 'I give myself up entirely to Coptic...I wish to know Egyptian like my French, because on that language will be based my great work on the Egyptian papyri.'

A month later he stated his position: 'I only dream Coptic and Egyptian...I am so Coptic that to amuse myself I translate everything that comes into my head...It is the true means of putting pure Coptic into my head. After that I will attack the Papyri and thanks to my heroic valour, I hope to come to an end. I have already made a great step.' He now began to tackle the demotic inscription on the Rosetta Stone once again, modifying the demotic alphabet that he had worked out the previous summer and declaring that he wanted to continue research into papyri: 'The papyri are always before my eyes – it is a prize so beautiful to win. I hope it is my destiny ...'

Despite his fright over Lenoir's publication, Champollion was still so over-confident that criticism of previous failed attempts at decipherment came easily to him. Of Åkerblad, he said that 'he was not able to read three words consistently in an Egyptian [in this case demotic – the term 'Egyptian' was used loosely for the demotic, hieratic and occasionally Coptic scripts] inscription,' and of Zoëga, who had a little earlier compiled the extensive corpus of hieroglyphic signs without deciphering any, he complained that he 'assembled an extraordinary quantity of material for a vast building...and did not set stone upon stone!' Likewise, he felt that Palin's work on the hieroglyphic script of the Rosetta Stone was useless, and that it was necessary to start the decipherment process from scratch rather than use any previous misguided research. The material that Champollion had at his disposal for this study was incredibly limited, and in a letter to his brother he listed all the hieroglyphic or hieratic (described by him as 'cursive') inscriptions on papyri or on mummy bandages that he knew were copied out in full or in part, in France and elsewhere – only seventeen in number, although he knew that a few others were awaiting publication, and he had seen some of the drawings of monuments with their hieroglyphs recorded by the savants in Egypt.

In June 1809 the very first volume of the *Description de l'Égypte* was published, presenting a mass of information compiled by the savants during Napoleon's expedition. Included were descriptions of Egyptian zodiacs, such as that at Dendera, as well as ancient monuments at many other sites in Upper Egypt, with hieroglyphs depicted on some of the accompanying plates. On seeing it, Champollion praised the fact that there were 'numerous Egyptian manuscripts, engraved with an astonishing accuracy, in this magnificent collection, as well as the prints, the drawings and the engravings, which alone have been able to serve as the solid foundation for the research of archaeologists,' but he went on to criticize the savants' attempts, led by Jomard, at decipherment: 'I don't have a great respect for them, they will be able to give us rather good drawings, but their explanations are really *water from the black pudding.*' Arrogant though it was, his assessment was accurate, and although his criticism could be harsh, it was usually expressed privately in letters to his brother.

For four months now, much of the work on his book *L'Égypte sous les pharaons* – his study of the geography, history and language of Egypt – had been complete, although the difficult compilation of an accurate map of the Nile Valley with place names was still not finished. This project was not helped by the fact that for strategic military reasons the precise surveys and maps produced during Napoleon's Egypt expedition remained a state secret and were not published for nearly another twenty years. Jacques-Joseph advised him to publish straightaway, but while he was considering this task, Étienne-Marc Quatremère published in June 1809 *Recherches critiques et historiques sur la langue et la littérature de l'Égypte* (Critical and historical research on the language and literature of Egypt). A former student of de Sacy, Quatremère was now working in the National Library and was described by Champollion as an envious person and nothing more than an egoist. Quatremère's work was a history of the Coptic language in Egypt and its early studies, but Champollion regarded this as his own field of research, forming the very foundations of hieroglyphic decipherment, and in his disappointment he was jealous that his professor, de Sacy, enthusiastically greeted the publication of this rival work.

The reaction of Champollion was to devote himself ever more to his Coptic studies, in order to do further research for his own book, itself partly based on evidence from Coptic manuscripts. Jacques-Joseph, who was increasingly anxious for his brother to pursue his hieroglyphic studies more vigorously, urged him to translate the Greek text of the Rosetta Stone into Coptic in order to see if this would provide the key to the demotic script and in turn the hiero-glyphic text. After some consideration Champollion wrote to his brother that this suggestion was impossible, because the names in the demotic inscription that Åkerblad had identified seemed to occur in different positions in the Greek text. He therefore felt that either the word order in the demotic was totally different from the Greek version, or that the demotic was not even the same inscription. Like Åkerblad and de Sacy before him, he did not even begin to look at the hieroglyphic inscription.

In the summer of 1809 his studies at Paris were at an end and his dreams of deciphering hieroglyphs unfulfilled. Jacques-Joseph arrived in Paris in August to help him travel home, and Champollion learned that he had been promised a position on the staff of the university faculty of history which was soon to be created in his beloved Grenoble. The future was looking very promising, when he was suddenly threatened with conscription yet again. By now 200,000 French troops were dying each year in the war in Spain alone, and the Emperor Napoleon was forced to conscript younger and younger recruits. The chances of survival for new recruits were extremely low; ten per cent of them deserted, many others hid from the draft or mutilated themselves to make them unfit for service, and there was an increase in marriage as only bachelors were eligible for conscription. It was within the power of his professor Louis Langlès to prevent his conscription, but still bitter at Champollion's refusal of the position of Consul in Persia, he did nothing. Yet again it was through the intervention of Fourier, Prefect of the Isère, that the threat of conscription was averted – for the time being.

# CHAPTER FOUR

# (The Teacher)

After two years studying at Paris, Jean-François Champollion returned to Grenoble on 15 October 1809, appointed as a teacher in a university that did not yet exist. As part of Napoleon's reforms of the education system in France a national network of universities was being created, and Champollion had become joint Professor of Ancient History at the one that was due to open in Grenoble. It was an acknowledgement of his remarkable ability that he should be given such a post at just eighteen years of age. Jacques-Joseph, then thirty years old, had left his cousins' mercantile business a few months earlier in July, having had the equal good fortune to be made Professor of Greek Literature and Secretary of the Faculty of Literature. He was already the assistant librarian of the municipal library at Grenoble, which was one of the first towns in France to have acquired such an institution, set up in 1772 by a public sub-scription that enabled it to purchase a huge collection of 34,000 volumes, the property of a local bishop who had just died. The library, along with a small museum, was situated on part of the first floor of the *lycée* so hated by Champollion, and at its heart was the Académie delphinale, the learned society of which Jacques-Joseph was still secretary. Jacques-Joseph also now had a second child who had been born in July – a daughter called Amélie-Françoise.

In the first few months back in Grenoble, Champollion was more concerned with preparing his university course than with pursuing his own research, but in March 1810 Jacques-Joseph received a completely unexpected letter from Silvestre de Sacy discouraging his former student Champollion from continuing research on hieroglyphic decipherment. Although at the time it appeared inexplicable, his action was undoubtedly dictated by his unwillingness to be eclipsed by the brilliance of his own students, several of whom he would denigrate in the coming years. Asking him not to abandon Oriental literature, de Sacy had written, 'I don't think he should get attached to the decipherment of the inscription on the Rosetta Stone. Success in these sorts of researches is rather the effect of a happy combination of circumstances than the result of persistent work.' If he thought such advice would be heeded, de Sacy really did not know Champollion, who wrote shortly afterwards to his good friend Antoine-Jean Saint-Martin in Paris, until recently a fellow student, that he was eager to resume work on the hieroglyphs despite what de Sacy had written.

The university at Grenoble opened in late May 1810. Napoleon had issued a decree that the post of professor in the new universities was equivalent to a doctorate, and so four months earlier both Champollion and Jacques-Joseph had received these degrees. Because of his age Champollion was only entitled to a salary of 750 francs per year, half of what he had received while a student in Paris. The older professors appointed at the same time received 3,000 francs, but because all the professors began at the university on an equal footing, with nobody having seniority, some of Champollion's colleagues were immediately jealous of his success. The other joint Professor of Ancient History was the elderly Jean-Gaspard Dubois-Fontanelle, who was also Dean of the Faculty of Literature and the librarian of the municipal library, where he was assisted by Jacques-Joseph, who was in turn assisted by Champollion. Dubois-Fontanelle generously gave up half of his professorial salary to Champollion, raising his earnings to 2,250 francs.

During the months leading up to the opening of the university, Champollion worked hard on preparing a meticulous course of history lectures. In setting up the universities, Napoleon had decreed how the professors should teach: to communicate the facts without comment and criticism, philosophy or oratory and to devise effective examinations for the students – in short, not to make any observations which showed the political regime was anything less than perfect. This approach clashed with Champollion's ideal of intellectual freedom, and from the start his lectures ran close to or beyond the limits set by the Emperor. These lectures, which caused a sensation, were based on the association of geography and history, criticism of sources and his favourite themes of chronology and the origin of mankind – dangerously close to a direct challenge to the Church which believed the world was no more than 6,000 years old.

Just when Champollion was starting his career as a university professor, Josephine's short reign as empress was coming to an end. She had many enemies among Napoleon's immediate relatives, most of whom had been against the marriage and had taken every opportunity to urge Napoleon to divorce her. Still in love with his wife, Napoleon did not want to take this step and would not have done so if Josephine had been able to give him a son and heir, but the marriage was childless and Napoleon needed children to inherit his empire. He finally announced the divorce in November 1809, and the marriage was annulled two months later. A suitable political marriage was then arranged, and in April 1810 Napoleon married the daughter of the Austrian Emperor – Marie-Louise, Archduchess of Austria. Although never as popular with the French as Josephine had been, Marie-Louise was accepted as empress, and opposition largely disappeared after she gave birth to a son the following year. The boy was given the title 'King of Rome' in the expectation that he would one day rule a united Italy, and everyone was relieved that a dynasty had been established. Napoleon was so optimistic that when Madame de Schwartzenberg, wife of the Austrian Ambassador, came to congratulate him on the birth of his son, he gave her a stone scarab covered in hieroglyphs that he had obtained during the Egypt expedition and had carried with him ever since: 'I have always worn it as a talisman,' he said. 'Take it, for I have no need of it now.' In fact he needed it more than ever, for his empire was just starting to collapse.

For a short while the events affecting the French Empire did not impinge on Champollion's life as much as the enmity of some of his colleagues and students. He was teaching some young men who had been his fellow students at the *lycée* and who now resented his elevation to professor. More serious was the jealousy of some of the other professors who had been teachers at the *lycée* when he was a pupil there: a few even complained directly to the Emperor that they had been passed over in favour of Champollion and his brother who were younger and less experienced – a complaint that Napoleon curtly dismissed. Characteristically, Champollion struck back at his enemies within the university with ridicule, composing a satire, the *Scholasticomanie*, which was successfully performed in the salons of Grenoble. Although he did not put his name to the work, some of his enemies who were the butt of his wit guessed who was responsible and became more determined in their opposition.

Champollion's preoccupation with academic politics was shattered just two months after the opening of the university, when he was thrown into a panic by the arrival of an order to present himself at an army depot within forty-eight hours, ready to take his place among troops being sent as reinforcements to the war in Spain. After the French victory over Austria at the Battle of Wagram in July 1809, the main area of conflict was now Spain, where the resentment of the French troops was expressed in the graffiti they left on the walls of the Spanish towns they were forced to evacuate: 'Spain: Generals' Fortune, Officers' Ruin, Soldiers' Death.' Elsewhere in Europe there was relative peace – the calm before the storm.

Yet again the Prefect, Joseph Fourier, came to the rescue, using his influence to obtain exemption for Champollion on the grounds that he 'intended to enrol in the Normal School,' whose students still had automatic immunity from conscription. Champollion never took up a place at the Normal School, but his ongoing 'intention' to do so proved sufficient to prevent any further attempts to force him into the army.

Once settled into the routine of a university professor, he again turned his attention to his research on Egypt, although no longer with the benefit of having at his disposal the collections of material that were available in Paris. The problem of decipherment remained a daunting one, and it was exemplified by the texts on the Rosetta Stone: the text of the Greek script was written in the Greek language and could be understood, but neither the hieroglyphs nor the demotic script could be read, and nobody knew what language or languages the texts of these two scripts were written in. To read the hieroglyphs it was necessary not only to understand how each hieroglyph functioned (whether it represented a sound, an abstract idea or was merely a picture of what it represented), but also how it related to the language it was being used to record. Was it, for example, like the European alphabet that could be used for writing many different languages, or was it like the Chinese alphabet whose hundreds of pictograms can only be used for Chinese and some closely related languages? The relationship, if any, between the hieroglyphic, demotic and Greek scripts was also unknown. With so many possibilities and so few clues it was not so unreasonable for de Sacy to assume that when it was achieved, decipherment was likely to owe more to good luck than hard work.

The difficulty of the problem would be reflected in the number of Champollion's hypotheses, the frequency with which he would change his mind as he enthusiastically embraced first one theory and then another, and with hindsight the fact that most of his initial ideas were completely wrong. Hundreds of different hieroglyphs had so far been identified but not deciphered, presenting researchers with the problem of how to account for such a large number of symbols. In early August 1810 Champollion, only nineteen years old, gave a lengthy talk to the Académie delphinale outlining his latest ideas on Egyptian writing, in which he refuted the theories propounded by earlier scholars such as Kircher and Warburton, and particularly those linking Egyptian with Chinese. Greatly influenced and so to some extent led astray by ancient Greek writers on Egypt, Champollion suggested that there were four types of Egyptian writing – one used for everyday purposes and business (demotic) and one used for writing sacred liturgies (hieratic), while the other two were types of hieroglyph: one type used for monumental inscriptions and one type for secret symbolical writing used only by the priests. Although a long way from the truth, he did realize that demotic and hieratic were two different cursive scripts, but he failed to understand their relationship with hieroglyphs. At this stage he suggested that hieratic and demotic were composed simply of letters of the alphabet and only differed from each other by the grouping of their signs, and that a more complex hieroglyphic system had developed from these two scripts, with signs that represented ideas not sounds, except in the case of foreign names. Champollion did at least get away from the centuries-old theory that hieroglyphs were entirely symbolic or ornamental – a set of mystical symbols devised to hide sacred information from all but a few initiates – rather than a writing system to communicate information.

Champollion's problem, now and in the future, was that although he worked out many theories he was too cautious, holding on to his findings and not publishing anything until he was absolutely convinced of his facts or until he was forced to publish by some crisis – giving a lecture to a provincial academy at Grenoble was not the way to stake his claim to any advance he had made towards decipherment. Only a crisis involving Étienne-Marc Quatremère, the former student and favourite of de Sacy, forced Champollion to begin publishing his book on the geography and history of Egypt, a work that had been started three years before in Grenoble, continued in Paris, and which his brother had been advising him to publish for over a year.

Quatremère, who had been working in the National Library at

Paris, had recently gone to Rouen to take up the professorship of Greek language and literature at the university (from where he would return to Paris only two years later in 1811). In June 1809 he had published his work on the Coptic language, to the annoyance of Champollion who had then felt obliged to further refine rather than publish his geography book, which he now called *L'Égypte sous les pharaons ou recherches sur la géographie, la langue, les écritures et l'histoire de l'Égypte avant l'invasion de Cambyse* (Egypt under the Pharaohs, or research on the geography, language, writing and history of Egypt before the invasion of Cambyses). To his horror Champollion now found out that Quatremère was preparing a rival work, pursuing a similar line of research to his own – studying the Coptic names of Egyptian towns and villages and deducing their original ancient Egyptian names (which at the time would have been written in hieroglyphs and hieratic). He knew that his own work on the geography of Egypt was more thorough and more accurate than that being prepared by Quatremère, as his sources of reference included Coptic and Arabic manuscripts, Greek and Latin texts, and accounts and maps by more recent travellers to Egypt, but until it was published nobody else could judge it. Knowing that publication of Quatremère's research was imminent, Champollion consulted with Jacques-Joseph, and they decided to try to steal a march by publishing the first part of Champollion's work as an 'Introduction.' Thirty copies of the sixty-seven-page work were printed in October 1810, but publication was not under their control and was delayed.

It was therefore Quatremère who won the race to publish, and January 1811 saw the appearance of his two-volume rival work, *Mémoires géographiques et historiques sur l'Égypte et sur quelques contrées voisines* (Geographical and Historical Reports on Egypt and Some Neighbouring Lands). Champollion's own 'Introduction' was not published until two months later, but, dismissive of his rival, he wrote in February to his friend Saint-Martin in Paris that he himself gave the names of 174 towns whereas Quatremère had only given

104. In the backbiting world of Parisian academics, a whispering campaign accused Champollion of plagiarism, when in fact neither he nor Quatremère had copied from each other, and very obvious differences of approach were apparent in their publications. Such accusations were what Champollion and Jacques-Joseph had expected and feared, but Champollion was particularly hurt and angry when he learned that his old tutor and supporter de Sacy, disliking the idea of the premature publication of the 'Introduction,' had convinced himself that the charge of plagiarism was true. Friends of Champollion in Paris, including Saint-Martin, tried to counter the accusations, but made matters worse by their exaggerated praise.

Later that year de Sacy published a review of the 'Introduction,' supporting Quatremère at the expense of Champollion, and as far as Champollion was concerned, war had broken out between himself and de Sacy and Quatremère. Jacques-Joseph was so upset at the attack on his brother that he wrote an admonishing letter to de Sacy, who did his best to defend himself, replying that: 'I have never had a keener desire than to encourage the young people, especially those who have sufficient courage to devote themselves still today to scholarship and to follow a career which offers more thorns than it promises crowns...I have not even had the thought of suspecting M. Champollion of plagiarism.' Despite his reluctance to publish material on the Egyptian writing system, the 'Introduction' to his L'Égypte sous les pharaons was far from being the first work that Champollion had published, as he had already contributed articles to academic journals and newspapers and produced various reports for Jacques-Joseph and Fourier for their contributions to the Egypt expedition's Description de l'Égypte, taking the realistic view that 'It is a harsh profession, that of an author.'

Although immersed in his research, he could not remain oblivious to events taking place outside the university as all too often they had an effect on his own life. A rift was growing at this time between Fourier and the two brothers, part of the reason being the embarrassment caused to the Prefect by their involvement with the newspaper *Annales du Département de l'Isère: Journal Administratif* (Annals of the Department of the Isère: Administrative Newspaper) which Jacques-Joseph had edited since 1808, and for which Champollion occasionally provided contributions. Imperial censorship of all publications was very strict, and on several occasions Jacques-Joseph had found his articles for the newspaper being censored, to the discomfort of Fourier. At the heart of the rift, though, was the brothers' involvement with the preface to the *Description de l'Égypte*, because Fourier had made no acknowledgement of their contribution to its preparation. While this was undoubtedly a disappointment, they were more aggrieved by reports from friends in Paris that Fourier had denied they had any involvement whatsoever. Fourier took action to distance himself from them both, avoiding them as much as possible and, in contrast to the previous friendly relations, dealing with them on only the most formal basis. Jacques-Joseph was later to write of this period: 'at this time my relations with Fourier were less intimate, less frequent; someone had come between him and me.'

Fourier seems to have been convinced by someone that close involvement with the two brothers was dangerous: probably by Ambroise Lepasquier, secretary to the Prefect, who was jealous of Jacques-Joseph. One of the more influential enemies of Champollion and his brother, Lepasquier had become even more embittered on learning that Fourier preferred Jacques-Joseph and had on several occasions tried to persuade him to take the post of secretary in Lepasquier's place – something Jacques-Joseph had always refused. For Fourier to be persuaded to abandon old friends is not so surprising, since this was a time of increasing political intrigue and unrest in France. Conscription made more and more demands on the people, businesses were disrupted because men had been conscripted, and others were in hiding to escape the draft or had returned from the wars disabled. Many families were in mourning for the loss of fathers, sons and brothers, and a general belief in the invincibility of Napoleon no longer prevailed.

By 1812, when Napoleon set out to invade Russia following the collapse of his alliance with that country, there was massive unemployment and soaring inflation. Hundreds of businesses went bankrupt, and while the rich could afford to eat, imported goods were only available on the black market at extortionate prices: tens of thousands of the poor were dying annually from malnutrition. On top of all this misery, the funds levied from the occupied territories no longer covered the cost of Napoleon's armies, forcing him to impose a war tax on the French. The invasion of Russia turned out to be a disaster, and of the 600,000-strong army that Napoleon led to Moscow, fewer than 100,000 men returned, most of the casualties dying from cold, hunger and disease. The losses of the Russian campaign would set off further rounds of conscription: France was bled dry of men and boys, and the people reacted against Napoleon's continuous wars, making it difficult for recruiting parties to collect the conscripts. Gangs of deserters and reluctant conscripts terrorized an increasingly lawless France where even the mounted gendarmerie had been conscripted into the cavalry. Despite the censorship and the propaganda, it was becoming obvious to the French that the empire was about to collapse, and Royalist factions were plotting to replace Napoleon by a king.

In Grenoble the death in February 1812 of Jean-Gaspard Dubois-Fontanelle, Champollion's co-professor of Ancient History, robbed him of the extra money that Dubois-Fontanelle had provided, and he reverted to his meagre official salary of 750 francs. Jacques-Joseph took over from Dubois-Fontanelle as municipal librarian on a salary of 900 francs, with Champollion as his unpaid assistant, and he also applied for the post of Dean of the Faculty of Literature, while Champollion applied to be full Professor of Ancient History instead of joint professor. His jealous enemies said he was too young and criticized his politics, calling him a 'Jacobin,' an insult implying that he was not only against the monarchy, but also anti-Napoleon. The Jacobins, deriving their name from the rue St. Jacques in Paris, where they had first met, became the most extreme democrats of the Revolution, and the name came to be applied to anyone holding similar extreme political views. For Champollion, this was only a slight distortion of the truth, since he was an idealist, not a supporter of Napoleon nor in favour of a monarchy, but held true to his belief that a democratic republic was the only fair government for France.

The post of Professor of Ancient History was not automatically given to Champollion but was put up for open competition, and his opponents tried to arrange for the chairs of Ancient History and Greek Literature to be combined so that if an outsider was appointed both he and his brother would lose their posts. The rift between the brothers and Fourier now had an effect, because Champollion and Jacques-Joseph were generally seen to be out of favour with the imperial authorities. This situation was made worse because as editor of the newspaper *Annales de l'Isère* Jacques-Joseph had published articles, without submitting them to the censor, about the terrible losses that the imperial armies had suffered in Russia and Spain. Despite these losses being common knowledge, the opportunity was taken while Fourier was in Paris to dismiss Jacques-Joseph from the editorship for publishing 'harmful assertions' – a dismissal that Fourier could do no other than confirm when he returned to Grenoble. Jacques-Joseph not only lost the editor's salary, but he also lost prestige since Fourier was obliged to report the matter to the government.

With so much at stake, Jacques-Joseph decided to go to Paris in September so as to put his own and his brother's case directly to the ministers responsible for education, only to find that the complaints and slanders of their enemies had even reached the government there. Nevertheless, he was successful in presenting his side of the dispute, returning triumphant to Grenoble, and soon after he was confirmed as Dean of the Faculty of Literature and remained Professor of Greek Literature, while Champollion was appointed Faculty Secretary and kept his post of 'temporary' Professor of Ancient History. The chair of Ancient History was left open to competition, but

as nobody wanted to apply, Champollion remained in the post and
Jacques-Joseph obtained a promise that the salary would rise to
2,250 francs.

Through these months of worry and financial hardship Champollion had made little progress in his research into the writing
systems of ancient Egypt, not helped by the lack of source material
available to him in Grenoble, which forced him constantly to beg
copies from other people. Research of a different kind, an experiment he conducted towards the end of 1812 on a Canopic jar, did
prove very successful. So-called Canopic jars of pottery or stone,
with their lids in the form of four heads of Egyptian deities, had been
discovered throughout Egypt and were at that time thought to
share the same function and date as entire jars (not just the lids)
moulded in the form of human heads that were found only at the
port of Canopus. The type of jar found only at Canopus was once
believed to have been worshipped as a personification of a god also
called Canopus, but later research has shown that these jars were
in fact revered as personifications of the Egyptian god Osiris and are
Greek or Roman in date. From his extensive research on the geography of ancient Egypt, Champollion knew that Canopus was unlikely
to be the ancient Egyptian name for the port, and he had also found
no evidence for a god called Canopus in ancient Greek and Latin
texts. The port was actually only named Canopus in Greek times –
known later as Aboukir, it had witnessed the landing of Napoleon's
expedition in July 1798 and the destruction of the French fleet by
the British a month later.

Even at this stage, Champollion realized that the two types of jar
must have had different functions and rituals, because the jars
shaped as human heads were restricted to Canopus and dated from
late Greek times. The misnamed Canopic jars with only their lids in
the form of heads were much more common and widespread, often
being found in tombs by Arabs who usually emptied them of their
contents before selling them to dealers and collectors. The little
museum that was part of the municipal library at Grenoble had two

such alabaster Canopic jars, twelve inches and sixteen inches in height, with lids that represented the head of an ape and the head of a jackal, and as the contents of the smaller one were actually still intact, Champollion decided to examine it to try to determine its function. The contents had formed a solid lump in the bottom of the jar, and so he took the drastic step of placing the jar in boiling water for half an hour to melt the original embalming fluid, from which he recovered an object wrapped in cloth. Once unwrapped, the object was shown to a naturalist from the Museum of Natural History at Paris who was passing through Grenoble at that time, and he confirmed Champollion's suspicion that the object was an internal organ that had been embalmed, probably a human heart, liver or spleen. The two jars are today on display in the Museum of Grenoble: the one with the ape-head lid is slightly blackened where the tar-like contents spilled over during the experiment.

From this work Champollion drew the conclusion that the four heads seen on the lids of the Canopic jars (woman, baboon, hawk, jackal) were those of four symbolical spirits, which according to the myths of the Egyptians presided at the examination of the soul before the tribunal of the god of the Underworld. Once hieroglyphs were deciphered, the results of this remarkable archaeological experiment were confirmed, and the guardian deities of the jars were identified as the 'Four Sons of Horus': the 'woman' was in fact (the god Imsety), guardian of the liver; the jackal was (the god Duamutef), guardian of the stomach; the ape was (the god Hapy), guardian of the lungs; and the hawk was (the falcon god Qebehsenuef), guardian of the intestines. A set of such Canopic jars was often used to store the embalmed internal organs which were removed during mummification. In the most thorough version of the process, only the heart and kidneys were left within the mummified body, while the brain was removed in pieces with a hook, usually inserted through the left nostril, and these pieces discarded. Most mummies were originally accompanied by a set of Canopic jars or by a set of 'Canopic packets' of embalmed viscera, which often had models of the guardian deities included in their wrappings. The actual process of mummification is one of the few aspects of ancient Egypt where hieroglyphs cannot help: there are no known accounts of the process in ancient Egyptian literature, and present knowledge is derived from descriptions by Greek and Roman writers, from examination of mummies, and by modern experiments. Champollion's modest experiment led the way in this field, but what he could not do was correct the misnomer 'Canopic jar,' which is still used by Egyptologists today.

Just before his experiment on the Canopic jars, with work continuing on the compilation of his Coptic grammar and dictionary, Champollion set out his latest ideas on the Egyptian language to his friend Saint-Martin in Paris – ideas that were beginning to change considerably. He no longer believed that hieroglyphs were later in date than the hieratic and demotic cursive scripts, although he still thought he could simply replace the cursive signs with their Coptic equivalents in order to be able to read and write ancient Egyptian texts. With regard to Coptic he stated that: 'I have so much analyzed this language that I strongly imagine teaching the grammar to somebody in a single day. I have followed its most perceptible chains. This complete analysis of the Egyptian language [Coptic] will incontestably give the basis of the hieroglyphic system, and I will prove it.' He would indeed prove it, but it would take many more years of work before he achieved his goal.

A few months later, in February 1813, Champollion revealed more of his ideas to Saint-Martin, but he still clung to the false notion that there was a secret priestly version of hieroglyphs. He now also identified two kinds of hieroglyphic sign, which was slight progress: 'In the hieroglyphs there are two sorts of signs: 1, the six alphabetic signs; 2, a considerable but defined number of imitations of natural objects.' Despite not being able to decipher a word of hieroglyphs, he began to make observations about word formation: 'The Egyptian nouns, the verbs and the adjectives do not have an ending, or rather a particular ending…everything is done by suffixes or prefixes.' Word endings, Champollion believed, were controlled by the alphabetical hieroglyphs – this was not absolutely accurate, but a step in the right direction.

In the university all had seemed well over the last few months, but the extra money promised to Champollion was not forthcoming and eventually, in February, it was confirmed that his salary was fixed at 750 francs, a quarter of the full professor's salary – his posts of Faculty Secretary and assistant librarian were unpaid. The situation suited the university administration which, like every other organization in France, was having to make whatever savings it could, but Champollion was depressed at being plunged once again into poverty, and consequently his health suffered.

By now the two volumes of Champollion's *L'Égypte sous les pharaons* were complete and had been entrusted to the Prefect Fourier to submit personally to the minister of education for publication. By April it was reported that the minister had reacted very favourably, but four months later Champollion found out that this was a deception and that Fourier had not even delivered the manuscript. After a long and heated meeting with Fourier, Jacques-Joseph believed he had repaired the poor relations, but as for publication he warned his brother 'We are constrained to wait, because the order for publishing must come from above.' Champollion's unhappiness increased because in July his sister-in-law Pauline Berriat died. Before going to Paris he had fallen in love with her and dreamed of marriage, sending an emotional letter to her from Paris via his cousin Césarine. Receiving no reply, he had eventually opened his heart to Jacques-Joseph, who was very sympathetic and did his best to comfort the rejected Champollion, but he was forced to admit that: 'Pauline was very angry, in consequence she laughed with Césarine at your letter and your thoughts.' Although the relationship had failed almost before it had begun, Champollion was still deeply affected by the death of Pauline, probably from tuberculosis, at the age of twenty-nine.

Because Champollion was so very depressed, Jacques-Joseph

took him off at the beginning of September to the remote Chartreuse mountains, north of Grenoble, where the Grande Chartreuse
monastery had been closed for two decades and was now the property of the nation. A contemporary guidebook for English tourists
in France recommended this area 'to those who are fond of seeing
the horrors of nature,' because they 'will be delighted with the way
leading to this convent, which is about six or seven leagues distant.
It abounds with steep mountains, impetuous torrents, irregular
rocks, frightful precipices, immense cascades &c.' During this trip,
the brothers, on behalf of the municipal library, successfully rescued from the former monastery nearly 2,000 manuscripts and
incunabula (books printed before 1500), which remain an enor-
mously important part of Grenoble's library collections today.

His physical and mental energies much revitalized after this trip,
Champollion set to work once again and was soon to write to Saint-
Martin with the startling pronouncement that he no longer
believed in the secret priestly hieroglyphs, which was a very important advance. By now Saint-Martin did not harbour a deep
indignation towards de Sacy for his treatment of Champollion, but
was beginning to support de Sacy's Royalist sentiments. Relations
between the two friends would eventually cool significantly, but
now Saint-Martin was sworn to secrecy over Champollion's new
discoveries, none of which he published because his views were
constantly changing during this period.

At the end of 1813 Champollion declared his wish to marry
Rosine Blanc, daughter of the glovemaker Claude Blanc. Glovemak-
ing was a highly respectable profession, virtually the only industry
in Grenoble, and Claude Blanc was one of the notables of the town.
He immediately refused to allow his daughter to be married to a
twenty-three-year-old professor, who had a bad reputation among
many of his colleagues at the university and who only earned 750
francs a year, with little prospect of advancement. Jacques-Joseph
was also against the marriage from the opposite viewpoint – he
thought that Rosine was intellectually too inferior for his brother.

With this refusal still ringing in his ears, the year 1814 began badly for Champollion, but it was much worse for Napoleon. After his disastrous Russian campaign of 1812, the Alliance of countries that was ranged against France began to close in: by October 1813 they had invaded southern France, and Napoleon also had to retreat from Germany. In December he was forced to abandon his struggle for Spain, and the Alliance threatened to penetrate farther into France if Napoleon would not agree a treaty. He refused to submit, and so in January 1814 the Alliance pursued their campaign deeper into France. Grenoble was threatened by the Austrians who had declared war on France the previous August (despite Napoleon's marriage to the Archduchess of Austria, Marie-Louise). People began leaving the town, and the remaining inhabitants organized themselves into a defensive force, of which Champollion and his brother were members. The seventeenth-century ramparts were patched up, weapons were collected, cleaned and loaded, and guards were posted. By now the defence of France, Grenoble, and above all the inhabitants, was all-important and along with everything else scholarly interests were put to one side, with Jacques-Joseph complaining to Champollion that they 'had other things to do than occupying themselves with hieroglyphs and the Egyptian language.'

At the beginning of April news arrived that the Austrian army was approaching, and although the town was in no state for a battle, let alone a siege, its zealous defenders were all for opening fire on the Austrians – couriers from Paris saved them from this action with the shocking news that the capital had fallen to the Alliance and that Napoleon had abdicated. An armistice was signed, and the Austrian army entered Grenoble. In May Napoleon went into exile on Elba, a small island east of Corsica, and the French monarchy was restored. Louis XVIII took the throne, with Champollion and Jacques-Joseph among the majority of the French population who were sorry to see Napoleon replaced by another Bourbon king. As a republican dreaming of a democratic France, Champollion had found fault with the Napoleonic regime, but to him a return to monarchy was a step in the wrong direction.

In Grenoble a minority of Royalists took the opportunity to make their presence felt, openly criticizing Fourier and even calling for his dismissal from his post as Prefect, as well as closing the town's theatre. Everyone was now forced to honour the new King, but with many people such 'honour' was merely lip-service, and those frequenting the salons often discarded the pretence of Royalism that they were forced to adopt in public. The theatre might be dark, but plays and satires were still performed for private audiences in the salons. Often subversive in their comedy, some of the plays were fearlessly and perhaps foolishly written by Champollion: usually parodies of stories from Classical Greek and Roman literature. They were apparently well received – even, on occasion, by officers of the occupying Austrian army, who enjoyed the entertainment untroubled by the ridicule of French institutions that gave the plays their satirical edge.

As the grip of the Royalist regime tightened throughout 1814, Champollion would come to view Napoleonic France, of which he had been so critical, as the lesser of two evils. He took to composing songs, set to popular tunes, for the private enjoyment of his closest friends. Needless to say, the songs were extremely disparaging about the Royalist regime and would have put their author in grave danger if he had been identified. So impressed were some of his friends by the songs that they had anonymous copies distributed throughout the town. People took to singing them late at night in the dark streets and alleys in order to taunt the police who were trying to suppress resistance to the new regime – there was soon an insatiable demand for these songs, which expressed so well their feelings of betrayal and defiance.

The mood of anxiety and disquiet in Grenoble was reflected in most other areas of France. In his last days of power Napoleon was seen as a hero battling to defend France against foreign invaders, in stark contrast to King Louis XVIII, who had been put on the throne by the invading foreign armies. The vast majority of the French army was still loyal to Napoleon, especially the thousands of troops returning to France who found there was little future for them now. The bourgeoisie and notables were equally concerned that prominent and lucrative posts were being reserved for Royalists, while nothing was being done for the unemployed, and the peasants were worried that the land they had bought during the Revolutionary confiscations would be returned to the aristocrats and that the old feudal obligations and taxes would be reintroduced. On top of all this, the national pride of the French was at a very low ebb since France had lost virtually all the territory it had acquired in over a decade of Napoleon's Empire. All the years of war and hardship appeared to have been for nothing, and as this had been brought about by the same foreign powers that had restored the monarchy, it was the King, not Napoleon, who was blamed.

In the university at Grenoble, as in many places, everything seemed to be in limbo as people waited to see what changes the new government would make. In May Jacques-Joseph went to Paris to try to find out what was happening, but by now Champollion was exhausted by the effects of his poverty, the responsibility of his three posts of Professor, Faculty Secretary and assistant librarian, and the crushing of his professional ambitions and personal desires. Despairing of the future he wrote:

> my fate is clear...I will try to buy a barrel [to live in] like Diogenes...I firmly believe that I was born at a bad time and that nothing that I want the most will ever succeed. I am pushed irresistibly by my head, my tastes and my heart into difficult paths, bristling with obstacles which are constantly replenished. Such is my destiny; it is necessary to suffer it no matter what ...

Diogenes, the Greek philosopher who founded the Cynic school of philosophy, was said to have adopted a life as near as possible to that

of the 'natural' life of primitive men, rejecting all material possessions, living in a barrel and begging for his food. He also rejected all
forms of education and culture, marriage, the family, worldly reputation and politics, and advocated sexual freedom. As Professor of
Greek Literature, the poignancy of this cry of anguish would not
have been lost on Jacques-Joseph, to whom it was addressed.

Amid all this turmoil, Champollion continued to use his friend
Saint-Martin as a sounding board for his ideas on hieroglyphs, as he
was still fearful of publishing them prematurely. He wrote at the end
of May 1814 that he continued to work on the Rosetta Stone
inscription, but in spite of some important findings, the results were
not as quick as he would wish: 'And the hieroglyphs? That's the
great question. I have many ideas, but I daren't boast of them before
obtaining some clear and solid success... one cannot guard against
oneself too much... I have already obtained a rather important
result... a hieroglyph alone, that is, an isolated hieroglyph, has no
value. They are arranged in groups.' While this is not strictly true,
since some hieroglyphs could meaningfully be used on their own,
Champollion was essentially right in that most hieroglyphs in a text
are arranged in groups, with each group conveying an idea. The
hieroglyphic phrase ⟨𓏞𓊹𓏏𓉐𓊹⟩ means 'I under-
took work in the temples of the gods,' and the hieroglyphs are
written in groups and arranged to make efficient use of the available space. The groups are:

| | | | | | |
|---|---|---|---|---|---|
| I undertook | work | in | the temples | of | the gods |

In June Jacques-Joseph arrived back in Grenoble with the news
from Paris that nothing would change within the university under
the newly restored monarchy. In the same month he was awarded
the Order of the Lily, First Class, which provoked scorn from Champollion who disliked such honours on principle and pointed out

that as so many had been awarded, it was almost a mark of distinction not to possess one. Outside the university, the Prefect Joseph Fourier was forced to put his Secretary Ambroise Lepasquier on unpaid leave, and his removal opened the way for a reconciliation between Fourier and the two brothers. It looked as if Champollion's *L'Égypte sous les pharaons* was about to be published, and he was given an increase in salary, though still only half of what his older colleagues received – certainly not enough to persuade Rosine Blanc's father to allow him to marry her. Also in June Jacques-Joseph was elected a corresponding member of the Institute of France's Academy of Inscriptions and Literature, and by August he was in Paris again as part of a delegation to assure the new King of the loyalty of Grenoble and to oversee the publication of Champollion's book. During his earlier visit to Paris, Jacques-Joseph had used all his diplomatic skills to ingratiate himself with the new regime, and he seized the opportunity to dedicate Champollion's *L'Égypte sous les pharaons* to King Louis XVIII, presenting him with a luxuriously bound copy of the work at a meeting on 12 August. On learning of Jacques-Joseph's intention just two days earlier, Champollion sent a furious letter to Paris to stop the dedication – it arrived too late, and Jacques-Joseph's diplomatic response to the new political situation triumphed over Champollion's unwillingness to hide his contempt for the monarchy.

By mid-September Champollion was looking forward to the publication of his book, but still very sensitive to criticism, he was fearful of the reception it would meet from adversaries such as de Sacy and Quatremère and wrote: 'Whatever happens, I will not lose for all that the satisfaction I feel for having brought to the world two big children who may well have some faults, but who at least give some hope.' At the end of October, his two volumes of *L'Égypte sous les pharaons* were officially published: the first volume was largely concerned with descriptions of various places in Upper Egypt and a discussion of their names, and the second was devoted mainly to Lower Egypt. In these volumes numerous tables of place names were presented, with Arabic, Coptic, Greek and possible ancient Egyptian equivalents, but the map of the Nile Valley that Champollion had struggled to compile was in the end confined to the Delta.

By now he was very dissatisfied with the copies of the Rosetta Stone at his disposal, as the inscriptions had still not been published by the Society of Antiquaries of London or by the savants in the *Description de L'Égypte*. With his own book just published and his confidence increasing, he decided that it would be appropriate to send copies to London, at the same time asking for clarification on parts of the Rosetta Stone text that were bothering him. On 10 November he wrote:

I have the honour of presenting you with the first two volumes of a work that I have undertaken on Egypt as it was before the invasion of Cambyses. The centuries which separate us from this period that is so important for the history of civilization have only left us with scattered and confused evidence of the ancient glory of this country. I have tried to gather the evidence together, and the volumes which accompany this letter are the first result of my work.

After describing his book in more detail, he went on to say that the language and writing of the ancient Egyptians was the most important element of his work and that he would by now have deciphered the Rosetta Stone if good copies of its inscriptions had been available to him:

The basis of my work is the reading of the inscription in Egyptian characters, which is one of the most beautiful ornaments of the wealthy British Museum; I mean the monument found at Rosetta. The efforts that I have made in order to succeed there have not been, if I may be permitted to say so, without some success; and the results that I have believed I have obtained, after a constant and sustained study, have made me

hope for greater things still. But I have found myself halted by
a difficulty that it is impossible for me to overcome. I possess
two copies of this inscription; one was done after the *facsimile*
that your Society had engraved, the other is the engraving of
the same monument which should form part of the third
volume of the Description de L'Égypte, published by order of
the French government. They present important differences,
sometimes small, sometimes sufficiently great to leave me in
an unfortunate uncertainty. Would I be permitted to beseech
the Royal Society to compare the transcribed passages on the
enclosed sheet, following the two engravings, with the monument itself. It is for me of the greatest importance to know
the true reading; and I am convinced that I would already
have determined the reading of the whole inscription if I had
had before me a plaster relief done by the most simple methods on the original; but having been forced to use two copies
which often give me quite varying versions, I am only going
step by step with an extreme distrust. One should not even
doubt that this essential part of ancient Egypt might today be
more advanced if a cast, as I have said, of the beautiful Rosetta
Stone had been deposited in each of the main libraries of
Europe: this new gift made to friends of literature would be
worthy of the zeal and selflessness which drive the Royal
Society.

In error, Champollion had addressed his letter to the Royal Society rather than the Society of Antiquaries which had first handled
the Rosetta Stone in London. It was the Foreign Secretary of the
Royal Society who replied to his letter. By sheer coincidence, Champollion had just made contact with his greatest rival in the
decipherment of hieroglyphs – Thomas Young – a rival of whom he
had been totally unaware.

# CHAPTER FIVE

# (The Physician)

Unknown to Jean-François Champollion, his most dangerous competitor was an Englishman who was already overtaking him in the search for a method of deciphering hieroglyphs. Thomas Young, who at the age of forty-one was seventeen years older than Champollion, had only become interested in decipherment a few months earlier, and his life and career were almost the opposite of the poverty, privation and political pitfalls that had so far hindered Champollion. Young had been born on 13 June 1773 at Milverton, then a small cloth-making town in Somerset. The oldest of ten children, his parents were both members of the Religious Society of Friends (Quakers), a Christian Protestant sect founded in England in the seventeenth century. He was brought up initially by his maternal grandfather Robert Davis, who encouraged this gifted child from the outset, constantly repeating to him: 'A little learning is a dangerous thing, Drink deep, or taste not the Pierian spring.' Young did drink deep and could read fluently at the age of two, he often memorized poetry, and started to learn Latin before he went to school. In March 1780, at the age of six, he was sent by his father to an inadequate boarding school near Bristol, and then to a similarly bad one in the same area: without good teaching, Young was forced to educate himself. He subsequently spent eighteen months studying at home until attending another school, at Compton in Dorset, from March 1782. Here he studied Latin, Greek, Hebrew,

French, Italian, mathematics, and 'natural philosophy' (as physics was then called), as well as several practical pursuits such as drawing and bookbinding.

Leaving the school four years later, Young continued his scientific and linguistic interests and was especially absorbed by Oriental languages: 'Mr Toulmin also lent me The Lord's Prayer in more than 100 Languages, the examination of which gave me extraordinary pleasure.' At the remarkably young age of fourteen he was employed as a private tutor and companion to Hudson Gurney, the grandson of the Quaker banker David Barclay of Youngsbury, twenty miles north of London. Here Young taught Gurney mainly Latin and Greek, while pursuing his own studies. He left his teaching post in the autumn of 1792, when Champollion was not even two years old, having equipped himself with considerable knowledge. Later he wrote 'whoever would arrive at excellence must be self-taught,' and the fact that they both outstripped their teachers and were forced to educate themselves is one of the few things Champollion and Young had in common. With these achievements behind him, Young went to London and began to study for the medical profession, in which a sound Classical education was then an essential prerequisite, with the encouragement of his uncle Richard Brocklesby, himself a distinguished physician who had two years earlier nursed his nephew back to health from tuberculosis. Through his uncle, Young had contact with many of the distinguished literary figures of the day, and as well as his medical studies he maintained his interest in languages, turning some of his ideas into short articles for magazines.

At the end of May 1793, only nineteen years old, Young read a paper to the prestigious Royal Society in London entitled 'Observations on Vision' in which he explained his discoveries relating to the anatomy of the eye, resulting in his election as a Fellow of the Society the following year, even though the surgeon John Hunter had immediately claimed the discoveries as his own and Young was accused of plagiarism. In October 1794 Young set out on horseback to travel nearly 400 miles to Edinburgh in Scotland in order to continue his medical studies, because the School of Medicine there had a very high reputation, and attracted students from many countries. Apart from circulating in society at Edinburgh, in his spare time he also managed to study German, Italian and Spanish, took dancing and flute lessons, and attended performances at the theatre. The Society of Friends would not have approved of such social activities, but before leaving for Edinburgh he had formally withdrawn from the Society, feeling no longer able to observe all its restrictions.

When the course of lectures in Edinburgh terminated in May 1795, Young left for a tour of Scotland equipped with more than forty introductory letters to the most distinguished, aristocratic houses. He returned to London in September and set out the following month for the university at Göttingen in northern Germany, where in November he continued his medical studies, while becoming increasingly fluent in the German language. The university at that time had one of the largest libraries in Europe, and drew students from across the Continent. Young's intention was to gain a diploma in spring 1796, then to amuse himself by travelling through Germany to Austria, Switzerland, northern Italy, and finally Rome and Naples, but his plans were thwarted by the war between Austria and France that was being fought on two fronts – in Germany and in Italy, where Napoleon Bonaparte was beginning to rise to fame as a general of the French army, two years before he set out on his Egypt expedition. Instead, Young left Göttingen in August 1796 and embarked on a limited tour of Germany.

In February the following year Young returned to England and the next month went to Emmanuel College, Cambridge, to study for a degree, having learned that he could not obtain a licence to practise medicine in London and its environs without two years' continued residence at the same university: his time at Edinburgh and Göttingen was deemed insufficient. While at Cambridge he met many eminent men, including Sir William Gell, who would later carry on a correspondence with him on hieroglyphs and other subjects. On 11 December 1797 Young visited his ailing uncle Dr. Brocklesby in London, just in time because Brocklesby died that very night, leaving his nephew his London house, library, pictures and about £10,000 in money: a substantial inheritance at that time. In stark contrast to Champollion, Young would never be short of money. In the spring of 1799, having completed the necessary six terms at Cambridge for his degree, he returned to London to commence his medical practice at 48 Welbeck Street, in a corner of the city that had until recently been surrounded by countryside known as Marylebone Fields. Now covering over thirteen square miles, London was rapidly expanding, its population of some 900,000 being twice that of Paris.

Before his medical practice became established Young accepted the post of Professor of Natural Philosophy at the newly formed Royal Institution in the autumn of 1801. The Royal Institution was designed as a school of science and Young was responsible for giving lectures and for editing its journal. Unlike Champollion, Young (by his own admission) was not a good lecturer, and after two years he resigned his post and devoted much of his time to preparing his lectures for publication. This massive work, in two volumes of over 700 pages each, was not published until 1807, and the promised sum of £1,000 was never paid to the author because the publisher went bankrupt shortly afterwards.

So that they could improve their French pronunciation, Young accompanied the two great-nephews of the Duke of Richmond to France for a month in the summer of 1802, during a short period of peace between France and Britain. While there, he visited Paris and went to meetings at the Institute of France – meetings also attended by Napoleon who was then First Consul and was spending most of his time in the capital dealing with affairs of state. Shortly after this excursion Young was appointed Foreign Secretary of the Royal Society based in Somerset House alongside the River Thames in London, a post which he held for the rest of his life.

On 14 June 1804 he married Eliza Maxwell, a marriage that was to be very happy, and their two sons, Robert and Thomas, were both to become physicians. For the next few years, while Champollion was being educated at the *lycée* and in Paris, Young undertook and published research relating to medicine and also practised as a physician, both in London and in the small town of Worthing on the south coast of England, where many of the wealthy and aristocratic would spend the summer: sea-bathing had become popular as a cure or relief for many ailments at a time when Continental travel to foreign health spas was restricted due to the resumption of war with France.

In January 1811, when Champollion had been teaching at the university at Grenoble for eight months, Young was elected one of the physicians to St. George's Hospital (now the Lanesborough hotel) at Hyde Park Corner in London. Just one mile from his home, St. George's was one of five new general hospitals founded by benefactors in the early eighteenth century. Young held this position at the hospital for the remainder of his life, but again he was not effective with the students: 'His manners were wanting in warmth and earnestness, and his want of knowledge of the difficulties of students made him pass over without sufficient notice the very points upon which they felt most desirous of being informed.' Fortunately for his patients he was a much better physician than teacher because he founded his practice on detailed observation. In 1813 his *Introduction to Medical Literature* was published, another large work like that of his lectures, and a new edition was brought out ten years later.

In the meantime, he undertook work in various other fields, being interested in many diverse subjects, but never applying himself to only one. His research tended to be published anonymously or under a pseudonym, as he did not wish to be seen to devote too much time to pursuits other than his profession of medicine. Even so, his authorship of most of these works was well known among scholars. In a letter to his friend and former pupil Hudson Gurney,

Young wrote: 'Scientific investigations are a sort of warfare carried on in the closet or on the couch against all one's contemporaries and predecessors; I have often gained a signal victory when I have been half asleep, but more frequently have found, upon being thoroughly awake, that the enemy had still the advantage of me, when I thought I had him fast in a corner,' prophetic words in view of his ensuing rivalry with Champollion and an indication that he often regarded his research as a process of triumphing over rivals rather than simply an advancement of knowledge.

Occasionally Young undertook research in Greek and Latin literature, notably on some of the 1,800 carbonized papyrus manuscripts which had been discovered at Herculaneum in Italy, one of the Roman towns that had been destroyed by a volcanic eruption of Mt. Vesuvius in A.D. 79. To the chagrin of scholars, these papyri turned out to be works of Epicurean philosophy and not long-lost works of ancient literature. In the *Quarterly Review* in 1810, Young had corrected the work of other scholars who had published their research on these papyri, and as a result his standing as an expert in this field increased dramatically. A few of the actual papyri were entrusted to the Royal Society, and he undertook experiments in unrolling them, although like many before him he generally only hastened their destruction. From the time of his work on the Herculaneum papyri, he was sent copies of inscriptions from near and far, mainly written in Greek or in Egyptian hieroglyphic or cursive scripts, for him to comment on.

Young's interest in Egyptian writing and language was particularly inspired by a papyrus written in hieratic which was found in 1811, wrapped with a mummy in a tomb near Thebes, by his friend the English traveller Sir William Boughton. Unfortunately the papyrus was soaked by sea water during its journey from Egypt and was in fragments when Sir William submitted it to Young for study in early 1814 – a time when Napoleon's French Empire was collapsing, when Grenoble was threatened with attack by Austrian forces, and when Champollion was exhausted by poverty and overwork.

Although coming to the Rosetta Stone somewhat late in the day, Young felt encouraged to embark on analyzing its three inscriptions during his annual visit to Worthing in the summer of that year. He equipped himself with the engravings done by the Society of Antiquaries of London and used the previous research of de Sacy and Åkerblad as his starting point. First of all he worked on the demotic inscription, to which he gave the name enchorial (from the Greek words *enchoria grammata*, 'letters of the country'), and a few years later he wrote with obvious resentment that his term enchorial was being overlooked, although he had been the first to give a name to this script on the Rosetta Stone:

> I have called these characters enchoric, or rather *enchorial*: Mr. Champollion has chosen to distinguish them by the term *demotic*, or popular; perhaps from having been in the habit of employing it before he was acquainted with the denomination which I had appropriated to them: in my opinion, the priority of my publication ought to have induced him to adopt my term, and to suppress his own.

It was the term demotic that became popular, and Young's terminology is now archaic.

Progress was slow and disappointing, and while Champollion was unaware of Young, so too was Young oblivious of the work that Champollion had been doing over the years, particularly since much of it was unpublished – the race to decipher hieroglyphs was taking place in a fog, with the competitors seldom glimpsing each other or even knowing if they were going in the right direction. Becoming curious to know what recent results had been achieved by de Sacy and Åkerblad, Young wrote to de Sacy in Paris in August 1814, who in late September confirmed that no real progress had been made, and revealed his disloyalty and hypocrisy towards not only Champollion but also Åkerblad, both former students of his:

M. Åkerblad has been at Rome for several years and although I have always been in correspondence with him and have often pressed him to make his results public he has never been willing to defer to my wishes. When he was in Paris he was not any more willing to communicate his work to me...I will not conceal from you that in spite of the qualified approval which I gave to the system of M. Åkerblad...I have always entertained grave doubts of the validity of the alphabet which he has made...I ought to add that M. Åkerblad is not the only one who flatters himself that he has read the Egyptian [demotic] text of the Rosetta inscription. M. Champollion, who has just published two volumes on the ancient geography of Egypt and who is active in the study of the Coptic language, claims also to have read this inscription.

In January 1815 Åkerblad wrote to Young admitting that he had done very little further work since his pioneering study of the demotic script of the Rosetta Stone, and only four years later he died suddenly in Rome at the age of fifty-five.

In his letter de Sacy was referring to the two volumes written by Champollion on the geography and history of Egypt (*L'Égypte sous les pharaons*), which had also been recently mentioned to Young by his friend Hudson Gurney, who was visiting Paris during the peace following the abdication of Napoleon in April. Now well aware of this rival, Young wrote to Gurney later that year: 'Notwithstanding what I have heard of Champollion, you will easily imagine that I am not a little anxious to see what *he has done*, and obliged to your kindness in procuring the book' – Gurney had sent Young a copy from Paris.

After a few months Young believed he had analysed the demotic text of the Rosetta Stone sufficiently to be able to arrive at a translation, and at the beginning of October 1814 he sent these results to de Sacy. By the end of the month he had also arrived at what he considered to be a translation of the hieroglyphic text, and although his

translations were mostly conjectural and incorrect, he had nevertheless surpassed the work undertaken previously by de Sacy and
Åkerblad and indeed published so far by Champollion. The word
'translations' is a misnomer for his results: what he had done was
divide the two inscriptions into groups of signs forming individual
words and phrases by applying his mathematical skills to the problem as if it were a code to be cracked. If the Greek, demotic and
hieroglyphic passages had been absolute literal translations of each
other it would then have been a relatively simple matter to match
words between the two unknown texts and the Greek text to arrive
at a full translation, but unfortunately this was not the case
because the scripts vary in their phrasing. Young's 'translations'
were anonymously published three years later in the Society of
Antiquaries of London's journal *Archaeologia*, as an appendix to his
report on Sir William Boughton's papyrus, 'though, for professional reasons, the discovery was made public with as little parade
as possible.' Young was very proud of his ground-breaking efforts,
and it cannot have been very welcome when he unexpectedly
received the letter from Champollion written in November 1814
asking for parts of the Rosetta Stone inscription to be checked, while
confidently implying that he could rapidly solve the problem if he
only possessed good copies of the inscription.

Only a few weeks later, at the end of February 1815, rumours
began to spread in France that Napoleon would come back from
exile to overthrow the monarchy, and in Grenoble some of his supporters received letters hinting at 1 March as the date of his
arrival – the town was being primed to receive him. On 4 March
news reached Grenoble that Napoleon had indeed landed three
days earlier and was marching on the town. By 7 March Napoleon
was at Laffrey, twenty-five miles to the south, where he met the first
resistance since setting foot in France: a detachment of men from
the fifth regiment stationed at Grenoble had been sent out to inter-
cept him. Napoleon's force of over 1,000 men could probably have
overwhelmed the detachment, but in order to win support he knew

that he would have to avoid setting French soldiers to fight other French soldiers, and there was a good chance that the men of the fifth regiment were sympathetic to his cause. Unbuttoning his grey greatcoat, he opened it to show his white waistcoat underneath and called out, 'I am here. Kill your Emperor, if you wish.' After a moment of indecision, he was met by a roar of '*Vive l'Empereur*' and the soldiers threw down their weapons, broke ranks and crowded round him, pledging their support.

In Grenoble everything was in confusion, with the inhabitants dividing into two factions for and against Napoleon, while a rowdy meeting of the magistrates and the Prefect Fourier argued about the best course of action. As Napoleon approached Grenoble on the night of 7 March it became obvious that the town would welcome him, and as he entered through the Bonne Gate for a triumphal ride through streets lined with people holding torches and shouting '*Vive l'Empereur*,' Joseph Fourier slipped quietly out of Grenoble, his position as Prefect compromised, even though he had once been a loyal supporter of Napoleon. Jacques-Joseph was among the crowd in the streets and later recorded, 'A magistrate standing beside me cried out as Napoleon passed by us, "Long live the Emperor, but long live Liberty!" "Yes," said Napoleon straightaway, turning his head towards us, "Yes, long live Liberty."'

With the troops at Grenoble joining him, Napoleon now had a credible army of over 8,000 men and thirty cannons. He remained in the town for a day and a half, during which time he asked the mayor to find someone who could help in his office. Recommending Jacques-Joseph, the mayor gave the old spelling of the family name, *Champoléon*, at which Napoleon exclaimed, 'It's a good omen – he has half my name!' – Jacques-Joseph was sent for and accepted the post. Napoleon spent many hours writing letters, despatches and orders, but also found time to meet various deputations, including the staff of the university. Jean-François Champollion was presented to him, and Napoleon remembered the name that had appeared on so many requests for exemption from conscription into the army. He asked about the work that was claimed to be so much more important than military service, to which Champollion replied that he had just finished his Coptic dictionary and grammar. Napoleon commanded Jacques-Joseph to bring the manuscripts to Paris where he would ensure they were published, commenting on the difficulties of publishing a Chinese dictionary which he had sorted out: 'they had worked on it for 100 years, but I had it done in three years through a decree.' The Emperor and the linguist, the two men who between them, but in such different ways, did the most to establish the study of Egyptology, met for just a few minutes – it was their only meeting.

The absence of Fourier was also noticed, and Napoleon was on the point of issuing an arrest warrant when Jacques-Joseph intervened, managing to convince Napoleon that he was not to be blamed, which resulted in Fourier being made Prefect of the Rhône and a Count of the Empire. At the end of March Fourier published a piece in a Lyons newspaper in which he said: 'I renew all my thanks to M. Champollion [Jacques-Joseph], and I beg him to be persuaded that I will never forget that he became closer to me when he believed my position threatened,' and speaking of his debt of gratitude to both brothers, contrary to his previous coolness towards them, he added: 'I will recall them always, in whatever situation that may be' – words reflecting the fact that the situation was constantly changing and unpredictable, because while Napoleon was meeting Champollion and Jacques-Joseph in Grenoble, representatives of Austria, Britain, Prussia and Russia were discussing how the Alliance against Napoleon could be reconstituted: the resulting treaty was signed at Vienna on 25 March.

Leaving Grenoble, Napoleon continued on his way to Paris where he arrived in late March, just as King Louis XVIII fled the capital for Ghent in Belgium. In the middle of April Jacques-Joseph risked everything by following Napoleon, leaving Champollion to shoulder his work, including the editorship of the newspaper *Annales de l'Isère: Journal Administratif*, a post returned to Jacques-

Joseph by Napoleon the month before. Jacques-Joseph took the manuscripts of the Coptic dictionary and grammar to Paris, where they were given to the Academy of Inscriptions and Literature for a formal assessment prior to publication. In Paris he worked on behalf of Napoleon, while in Grenoble the extra burden rapidly took its toll on Champollion's health, and he used an array of medicines and took frequent baths just to try to keep going – there were barely enough hours in the day to keep up with his tasks and virtually none to spare for studying hieroglyphs.

After all the upheavals of Napoleon's return from exile, Champollion finally received a reply from Thomas Young to his letter of the previous November, but what Young wrote regarding Champollion's problems with the Rosetta Stone inscription was hardly helpful:

I have had much pleasure and interest, Monsieur, in making the comparisons that you wished between the two copies of the inscription. In general that of the Society of Antiquaries appears almost perfect; sometimes, however, the French copy is the most exact: but in most of the places that you have cited there is some obscurity in the original features which are a little confused or worn, and it is only by comparing the various parts of the stone that one can be assured of the true reading.

He then went on to ask if de Sacy had passed on the translations that he himself had made of the Rosetta Stone, which must have been unpleasant news to Champollion, who did not know of Young's work. Young added: 'I do not doubt that the collective efforts of savants, such as M. Akerblad and yourself, Monsieur, who have so much deepened the study of the Coptic language, might have already succeeded in giving a more perfect translation than my own, which is drawn almost entirely from a very laborious comparison of its different parts and with the Greek translation.' To this letter Champollion replied in early May 1815, giving his thanks but stating that de Sacy had not informed him of Young's research.

By June, just as Jacques-Joseph was being made a Knight of the Legion of Honour by the new Napoleonic regime, the Royalists were beginning to recover, circulating propaganda news-sheets and attempting to levy taxes in the name of the King. In place of his brother, Champollion countered the propaganda as well as he could, once again composing satirical political songs that were extremely popular. On 18 June, as Napoleon was losing his last battle against the Allied armies of Prussia and Britain at Waterloo, Champollion published an article that was soon to have severe consequences for him and his brother. Attacking the legality of the Bourbon dynasty as rulers of France he wrote: 'There is no law of succession for the throne of France. The people alone award the crown; they gave it once to Hugues Capet [a king of France], and now they remove it from his descendants in order to entrust it to a more worthy person. The people's choice makes the only legitimacy. Napoleon is therefore our legitimate prince.' Champollion's decision to follow the lead of his brother and actively support Napoleon for the first time in his life was to prove disastrous.

Napoleon's return from exile was short-lived, because his abdication soon followed his defeat at Waterloo. The Royalists now had effective control of the government in France, and the Alliance forces began closing in. By 5 July Grenoble was under siege from a combined army of Austrians and Sardinians – it was the last town in France to offer any resistance to the invaders, and the main attack on the town began at six o'clock the next morning. Weapons had been distributed among the citizens, who had been organized to reinforce the garrison, and by ten-thirty the attack had been repulsed. The Royalists in Grenoble had assured the invaders that the town would immediately surrender, and so the Austrian and Sardinian army, surprised by the determined resistance, began heavy shelling. Champollion was on the ramparts before the attack commenced, but when the shelling began he was fearful for the library. Running through the rain of exploding shells, he climbed to the second storey of the library where he remained until the bombardment ceased, ready to put out any fires that threatened the irreplaceable collection of books and manuscripts.

With about 600 dead and 500 wounded, it was obvious to the attackers that it would be very costly for them to take the town by storm, so they proposed a ceasefire for three days, to which the people of Grenoble agreed. Jacques-Joseph was still in Paris and Champollion sent him an eyewitness account of the fighting, saying that the French cannon and the guns of the conscripts had smashed the Austrian forces who had withdrawn into the suburbs, and that Zoé (Jacques-Joseph's wife) had taken to the fighting 'like an Amazon.' After the announcement of a ceasefire, relief from the danger appears to have given way to a celebration of the temporary victory gained by the people of Grenoble, and Champollion declared that never in his life had he drunk such good ratafia.

By 8 July the Royalists had sufficient control over France for King Louis XVIII to return to Paris, and, seeing that further resistance was hopeless, Grenoble capitulated to the invaders two days later. In mid-July Napoleon left France for exile on the isolated island of St. Helena, a place which he had accurately summed up in one of his student notebooks many years before: 'St. Helena – small island.' By this time Champollion had sent two letters to his brother in Paris, blaming him for getting them both involved with Napoleon in such a highly visible way, but at the same time begging him to put all the blame on himself, because he had no wife and family to consider, but only his own skin to save:

Save yourself first of all. As for me, it will be whatever God wills. I gave my opinion because I believed it good and believe it good still. If they reproach you for the *jacobinism* of your newspaper, say boldly that it was *me who did it*, because that is true. If they need a victim, I am there. I have neither wife nor child...the important thing is that you get out safe and sound from the crisis.

Once again everything was in limbo in Grenoble, and Champollion despaired for the future: 'I try to resume my ordinary work, but my heart isn't in it. I am entirely despondent for the future and firmly believe that it isn't for me.'

At the same time Jacques-Joseph learned that the Academy of Inscriptions and Literature, swayed by the ardent Royalist de Sacy, had rejected Champollion's Coptic dictionary and grammar for publication, branding it not worthy of being printed at government expense. Utterly despondent and angry with de Sacy, whom he now considered his most mortal enemy, Champollion wrote to Jacques-Joseph an accurate prophecy of how political allegiance would soon carry more influence than anything else: 'I think that it is useless to continue a struggle which we must lose sooner or later. The spirit of the party will henceforth reign in France with more force than ever; the colour of the hat is going to decide the value of the ideas in the head that wears it. Everything is finished' – a perceptive prediction of the political and social change that was about to occur in France. He rose above his misery, despair and failing health for one last act of defiance. Before he lost control of the *Annales de l'Isère*, Champollion used it in his brother's absence in Paris to publish the 'Toast to the Republic,' which defined his political stance and damned him to proscription by the new Royalist regime:

> I have my loves, like many another
> But now I am forced to be specific
> I raise my glass and pledge forever
> All of my toasts to the Republic!

Silvestre de Sacy, such a devoted supporter of the monarchy, was by now very hostile to Champollion, who held very different political views, and he tried to discredit him outside France as well as within it. Later in July he told Thomas Young that he had passed on his demotic 'translation' (after many months) to Champollion's brother Jacques-Joseph. He also warned Young to be careful not to

trust Champollion, an extraordinary statement to make to someone in a foreign country with whom he had only exchanged a
handful of letters:

> If I might venture to advise you, I would recommend you not
> to be too communicative of your discoveries to M. Champol-
> lion. He would then be able to claim priority afterwards. He
> seeks, in many parts of his book, to create the belief that he
> has discovered many words of the Egyptian [demotic] inscrip-
> tion of the Rosetta Stone: but I am afraid that this is mere
> charlatanism: I may add that I have very good reasons for
> thinking so.

In his letter de Sacy was not even complimentary about his former
pupil Quatremère:

> You are not unaware, to be sure, that someone in Holland has
> also announced having discovered the alphabet of this
> inscription [the Rosetta Stone] and that in Paris M. Étienne
> Quatremère flatters himself as reading a large part. Whether
> I consider these discoveries real or imagined in theory, noth-
> ing appears less likely; for I hold it as certain that Coptic is
> nothing like ancient Egyptian, and the Greek translation
> seems to offer a sure means of decipherment: but as soon as I
> glance at the monument, I think differently, and I despair that
> one might ever finally read it.

Despite his pessimism, de Sacy congratulated Young: 'You appear to
have made great progress in the decipherment of hieroglyphs.'

Over the succeeding months, Young embarked on detailed correspondence with scholars such as de Sacy, Åkerblad and Jomard,
exchanging thoughts and ideas about hieroglyphs. Selected parts of
this correspondence were published by him in 1815 and 1816 in a
somewhat obscure journal, the *Museum Criticum*. As an enthusias-

tic supporter of Young, Edme Jomard had written to him recently in April 1815 that: 'The success you have obtained, Monsieur, in the interpretation of the Rosetta Stone has excited my interest and curiosity to the highest degree and, more than anything else, what you have done on the hieroglyphic inscription.' Young was anxious to extract data on hieroglyphs from Jomard, whose position as editor of the *Description de l'Égypte* gave him enviable access to large quantities of material gathered in Egypt. Young wanted this information so that he might compile some sort of hieroglyphic dictionary, but Jomard used the excuse that he was still working on everything, although he stated:

> It is a job that I undertake with a sort of affection, because I have realized that it would not be useless to the research of savants. I have always been surprised at the effort that so many skilful people waste in order to decipher hieroglyphs. It is as if one wanted to read and understand Hebrew without knowing the number or depiction of the letters of the Hebrew alphabet.

In reply, Young distanced himself from the scholars who Jomard thought were misguided and stated that he himself was trying to copy all the hieroglyphic inscriptions that came his way, but he still failed to persuade Jomard to share his information.

At the end of 1815, William Richard Hamilton (British Under-Secretary of State for Foreign Affairs) had good news for Young, because the newly appointed British Consul in Egypt, Henry Salt, had been instructed to purchase any examples of inscriptions deemed by Young to be important and to forward them to England. Over that winter Hamilton lent Young the volumes of the *Description de l'Égypte* that had been published so far. These volumes contained virtually no mention of the hieroglyphs themselves or the decipherment process, as Jomard still had ambitions to publish his own *Observations et Recherches sur les Hiéroglyphes* (Observations and Research on Hieroglyphs), but this work never materialized. Young started to analyze what had been published in the *Description de l'Égypte*, but much of it was of no use to him because of the rigidity of his approach. He was still looking to crack the code of the hieroglyphs, and, lacking Champollion's detailed understanding of ancient Egyptian culture as a whole, he was dismissive of the material: 'All the inscriptions on temples, and the generality of the manuscripts found with the mummies, appear to relate to their ridiculous rites and ceremonies: I see nothing that looks like history.'

Despite his restrictive methods, Young now reached a stage where he could make several important points about hieroglyphs, correctly realizing that some were pictorial (for example, 〚 means 'obelisk,' and 𓁹 means 'eye flowing with tears' – that is, weeping), while others performed different functions. Noticing how hieroglyphic plurals were constructed, he commented: 'in order to express a plurality of objects, a dual was denoted by a repetition of the character, but that three characters of the same kind, following each other, implied an indefinite plurality, which was likewise more compendiously represented by means of three lines or bars attached to a single character.' In this he was absolutely accurate. In ancient Egyptian there were three states: 'one,' 'a pair' and 'many.' The sign 𓊬 represents the word *neter*, meaning 'one god.' A 'pair of gods' was represented by doubling the hieroglyph: 𓊬𓊬 and the plural, 'gods,' was represented by three hieroglyphs: 𓊬𓊬𓊬. As Young also observed, the plural was sometimes abbreviated by using just one hieroglyph and three strokes, so 'gods' can also be written as: 𓊬𓏤 and 'houses' can be written as: 𓉐𓏤 . The plural adds 'w' to the word, so that *neter* ('god') becomes *neterw* ('gods') – strictly *ntrw*, since there are no proper vowels in ancient Egyptian – but because 'w' is pronounced 'oo,' it is often written as 'u' for convenience, so that 𓊬𓏤 is usually transliterated as *neteru*, and the plural of 'house' (usually transliterated as *per*) is *peru* ('houses').

Although Åkerblad had made minor comments on the Rosetta

Stone hieroglyphic numerals many years earlier, Young was now able to write: 'definite numbers were expressed by dashes for units, and arches, either round or square, for tens.' Jomard published an important work on hieroglyphic numerals in 1816, although Young believed Jomard had plagiarized his own work, which was probably not the case. The numerals actually work in a simple way, constructing any number by combining and/or repeating one or more of seven basic signs. These signs are:

| | (one) | ∩ | (ten) |
|---|---|---|---|
| ℮ | (one hundred) | ⌡ | (one thousand) |
| ⌠ | (ten thousand) | ⌐ | (one hundred thousand) |
| ⚮ | (one million or many) | | |

When combining the signs, the higher value signs are always written in front of lower value ones, so ℮IIII means 104, ⌡∩II means 1,022 and IIII/IIII means 9.

Up to now scholars, including Champollion, had assumed that demotic was entirely alphabetic (where the signs only represent single sounds which are combined to spell out words, as in English or ancient Greek). By examining the Rosetta Stone inscriptions and various papyri written in hieroglyphs and hieratic, Young now concluded that demotic (his enchorial) only used letters of the alphabet to spell out the sounds of foreign rather than Egyptian words, including words of the ancient Greeks, who were the last rulers of Egypt before the Roman conquest. Nevertheless, he could see that there were marked similarities between many symbols in his cursive enchorial script and the hieroglyphic script, and deduced from this that the cursive script characters 'are truly hieroglyphics, though in a corrupted form.'

Since Åkerblad's early identification of some of the names in the demotic script, progress on even the Rosetta Stone had been very disappointing across all of Europe. Young summed up the position:

> It was natural to expect that…the critics and chronologists of all civilised countries would have united, heart and hand, in a common effort to obtain a legitimate solution of all the doubts and difficulties, in which the early antiquities of Egypt had long remained involved. But, excepting Mr. Champollion and myself, they have all chosen to amuse themselves with their own speculations and conjectures: the mathematicians of France have continued to calculate, and the metaphysicians of England have continued to argue, upon elements which it was impossible either to prove or disprove.

He later defined the problem of decipherment as one requiring a lifetime's study:

> and when we reflect that, in the case of Chinese, the only hieroglyphical language now extant, it is considered as a task requiring the whole labour of a learned life, to become acquainted with the greater part of the words, even among those who are in the habit of employing the same language for the ordinary purposes of life, and who have the assistance of accurate and voluminous grammars and dictionaries: we shall then be at no loss to understand that a hieroglyphical language, to be acquired by means of the precarious aid of a few monuments, which have accidentally escaped the ravages of time and of barbarism, must exhibit a combination of difficulties almost insurmountable to human industry.

As all the rival researchers were aware, the greater the difficulty, the greater the prestige that success would bring.

While Young was enjoying the respect of the scholars of Europe, free to communicate and exchange ideas with them, in the paranoid climate of the restored monarchy, Champollion in Grenoble and Jacques-Joseph in Paris were placed under surveillance, and both were removed from their association with the local newspapers in Grenoble. In November 1815, shortly before Jacques-Joseph returned from Paris, Champollion wrote to him about a proposed change of career, abandoning everything for which he had worked so hard: 'No doubt the Faculty of Literature will be suppressed... We are razed to the ground... I wish to embrace the state of notary [lawyer] at Grenoble. You will tell me that it is priest become miller, but what does that matter if the priest has not eaten and they have flour at the mill?' Having been engaged to marry Rosine Blanc since 1814, Champollion hoped by this resolve to overcome the objections of her father, who regarded the law as the only secure profession in such difficult times. In deep despair at the way his world was collapsing around him and with all his hopes dashed, even his prospects of marriage, Champollion saw a change of direction as the only option and was determined to give up public education.

Despite their problems, Champollion and Jacques-Joseph still took the risk to help the fugitive General Drouet d'Erlon, who had fought for Napoleon at Waterloo and had been condemned to death by the new regime. They found him shelter throughout the winter of 1815–16, and the following spring, just a short distance from the prefecture, they aided his escape from France to Munich, where he joined Prince Eugène de Beauharnais, Josephine's son and Napoleon's stepson. It would have been extremely dangerous if the authorities had found out, but even without this there were already large dossiers on Champollion and his brother, whose enemies were pressing hard for their proscription. Worse was to come from changes in the university at Grenoble, as non-Royalists holding key positions were being replaced by Royalists, and in January 1816 the Faculty of Literature was closed down, as Champollion had predicted, robbing him and Jacques-Joseph of their academic posts. Attacks were now made on their involvement with the municipal library, and Champollion was accused of using it for political meetings, which he strenuously denied: 'It is absolutely false that there was ever held here any meeting other than that of the Academy of Grenoble.' Other false accusations against them both were lodged with the authorities and in March 1816 they were condemned to internal exile.

All Champollion's new hopes and plans for a change of career and his marriage were now shattered by the proscription order, and although the brothers used what was left of their influence to ensure they were exiled to their home town of Figeac, they were separated from their family and friends in Grenoble. Jacques-Joseph took his son Ali, but had to leave his other children with his wife Zoé – Jules, born in 1811; Aimé-Louis, born in 1812; and Zoé, born in 1815. Amélie-Françoise is no longer mentioned at this time and appears to have died very young. Champollion and Jacques-Joseph were only granted fifteen days for the journey, so they could not take too much baggage and had to leave most of their books and research notes in Grenoble. Passports issued for the journey provide rare personal details about the two men. Since their faculty at the university had been closed, Jacques-Joseph was recorded only as a librarian, thirty-six years of age, a native of Figeac living at Grenoble, travelling with his son who was six-and-a-half years old. He was described on the passport as 1.67 metres tall with dark brown hair, brown eyes and a light complexion. He had an oval face with 'medium' forehead and mouth, a slender nose and no particular distinguishing marks.

Champollion was recorded on his passport only as assistant librarian, twenty-five years of age, and 1.70 metres tall, with black hair, black eyes and 'brown' complexion. He had a large forehead, flat nose and a round and jutting chin, and his face was 'full round' and 'lightly marked by smallpox' – this passport is the only indication that he may have had this disease. In the section marked 'beard' on the passports, Jacques-Joseph's is entered as 'brown' and Champollion's as 'black,' but this does not necessarily mean they had beards at this time, as it may merely record the colour of their facial hair. Champollion later grew a beard when he was in Egypt, but most portraits of the brothers show them clean-shaven.

The passports were dated 18 March 1816 and cost two francs each. They specified that Champollion and Jacques-Joseph were to travel to Figeac via Lyons and Aurillac rather than allowing a shorter, more southerly route. This provision was for their own safety because the Royalist backlash that had broken out in southern France with the restoration of the monarchy was still continuing, and that region in particular was not under the full control of the new regime. The situation was complicated by the struggle between the different Royalist factions, which fell into two broad groups: the supporters of the reigning monarch Louis XVIII, who were liberal in their outlook and concurred with his view that it was impossible to return to the absolute monarchy that had prevailed before the Revolution, and the Ultras (extreme Royalists) who wanted to remove all the changes that had taken place during the Revolution and under the subsequent rule of Napoleon. While King Louis wanted to unite the monarchy with the people, the Ultra-Royalists supported his brother, the more reactionary Count of Artois. In southern France local groups of Ultras such as the *verdets* (wearers of a green cockade) had effective control of some areas and followed their own agendas while pretending to be on the side of the new regime. In some places religious bigotry led to the hunting down of Protestants, while in others companies of armed bandits threatened the safety of travellers. It was because of the threat of interception by such bands of outlaws and vigilantes, as much as the need to avoid areas in control of the Ultras, that Jacques-Joseph and Champollion were obliged to make a wide detour to the north, arriving in Figeac on 2 April 1816 after a journey of two weeks.

The brothers found that the town had changed little during the years they had been away, appearing as tranquil, if a little more prosperous, than they remembered, but their own family had not fared so well. Their mother had been dead for nearly ten years, and the family house in the rue de la Boudousquerie now only contained their father and two sisters, Thérèse and Marie, their other sister Pétronille having married in 1803. It was not a happy household: their father had become an alcoholic, causing a breakdown in his health, and the family bookselling business was in a desperate condition. Whether the cause or the result of their father's heavy drinking, the collapse of the bookselling business made it obvious that action was needed to prevent bankruptcy. The shop was now run by Thérèse who was forty-two years old, while Marie, who was thirty-four, kept house. Neither of them had married, perhaps because the failing business did not produce enough money for their dowries, and although they were both united in idolizing their brothers, the lack of money and worries about their father caused them to argue constantly about all other matters.

It seemed that Champollion and Jacques-Joseph had been forced out of one difficult situation into another, and at first both of them thought the little town of Figeac suffocating in comparison with Grenoble and Paris. Jacques-Joseph wrote to his wife Zoé that there were barely four or five people who could carry on an intelligent conversation, and that minutes seemed like days, days like months, and months like centuries. Champollion wrote to his close friend Augustin Thévenet (once a pupil at the *lycée* and now in charge of his family's shop in Grenoble) that he passed his time quite sadly, yawning and cursing all the day – he did not know when he would be able to see the beautiful mountains of Grenoble again. Yet they had left only just in time: in early May 1816 an uprising was led by Jean-Paul Didier, former director of the law school there. With a small force of ex-soldiers and peasants, many of whom were drunk, he raised a Napoleonic banner and tried to take control of the town. The 'uprising' was easily quelled, but the Ultras used it as an excuse for bloody repression, and shot eighteen of the participants. Didier fled, but was captured and brought back to Grenoble to be guillotined publicly. Many others within the town were under suspicion, and although already in Figeac, Champollion and Jacques-Joseph were implicated. Nothing was proved against them and no action was taken, but they might have lost their lives if they had still been in Grenoble.

While the two brothers found their home town initially depressing, they were regarded by some as celebrities and fortunately they were welcomed by the local Prefect, Count Lezay-Marnesia, once an officer in the French army in America and since 1815 Prefect of the Department of the Lot, which included Figeac. Passionate about archaeology, the Prefect persuaded them to look for the site of Uxellodunum, thought to be in the area around Figeac. According to the *Commentaries* of Julius Caesar, Uxellodunum was the last Gallic stronghold to be captured by the Roman legions. From 59 B.C. Caesar had systematically conquered Gaul, and he described Uxellodunum as being defended on all sides by steep cliffs, with a river running in the valley below, but with only a single spring for a water supply. His soldiers laid siege to the stronghold and dug a tunnel to divert the spring, so that the desperately thirsty Gauls were eventually forced to surrender. The Roman conquest of Gaul was then complete, and to deter uprisings Caesar ordered that all those who had fought against the Romans at Uxellodunum would have their right hands cut off. The French never forgot or forgave this conquest of their country, a conquest belatedly avenged by Napoleon's invasion of Italy, and so the site of Uxellodunum became an important symbol of patriotic resistance and the identification of its location was a correspondingly important project.

Jacques-Joseph, helped by Champollion, undertook the task of finding Uxellodunum, and through the summer of 1816 they searched the area trying to identify the place from clues contained in Caesar's text. They came to the conclusion that it was Capdenac-le-Haut, overlooking the River Lot and some three miles from Figeac – excavations conducted by Jacques-Joseph produced Roman finds which appeared to confirm the identification. The apparent success of the project greatly pleased Prefect Lezay-

Marnesia and enhanced the reputation of Jacques-Joseph and his brother among the educated elite of the area. Subsequent archaeological work has shown that the fortifications at Capdenac date to the Medieval period, and another site has been identified as Uxellodunum: a place called Puy d'Issolud, near Vayrac, approximately twenty-nine miles north-west of Figeac, where even the tunnel cut by the Romans to divert the spring has been found. Nevertheless, controversy persists, and today the people of Figeac maintain that Capdenac-le-Haut is the real location of Uxellodunum.

From that summer Champollion, as well as helping Jacques-Joseph look for Uxellodunum, also became immersed in completely reorganizing and revising his Coptic dictionary and grammar which had been rejected for publication, having managed to bring those precious manuscripts from Grenoble. Ultimately the importance of these works was that their compilation instilled in him such a thorough knowledge of Coptic, the surviving remnant of the ancient Egyptian language, that it gave him an immense advantage over most of his rivals in his approach to the decipherment of hieroglyphs, but in spite of all the efforts he made throughout his exile in Figeac, they were never to be published.

From the moment the two brothers were exiled, Jacques-Joseph began sending letters to Paris, begging friends and acquaintances to intervene with the authorities on behalf of his brother and himself. Such had been the change in the political climate that even old enemies of Champollion like Edme Jomard, editor of the *Description de l'Égypte*, and Louis Langlès, who had previously refused to try to prevent Champollion's conscription when he was teaching him Persian in Paris, now did their best to help. Of all Champollion's former professors, only the Royalist Silvestre de Sacy adamantly refused to intervene on their behalf, a refusal which damaged his reputation with some of the other scholars in Paris. More distressing was the behaviour of Saint-Martin, who had become an ardent Royalist, firmly aligning himself with de Sacy and distancing himself from his former friend Champollion.

Despite energetic lobbying by their friends and old enemies in Paris, as well as support from the Prefect in Figeac, it was only in November 1816 that Jacques-Joseph was officially released from nearly eight months of exile, and Champollion was not released until the following January. As the months dragged on, Champollion was free from the political intrigues of jealous colleagues and initially without an immediate goal in his life, his plans of becoming a lawyer having been wrecked by his enforced exile. In a kind of limbo, but regarded as a celebrity and popular with the local people, the stresses of the preceding years unravelled and he began to take pleasure in the tranquillity of Figeac. This was the calm before the next storm in his life, which began quietly enough through his brother's contacts in Paris. During a visit to England at the end of 1814 to record the monuments confiscated from the French in Egypt, including the Rosetta Stone, Jomard had been struck by the Lancaster method of teaching there and founded a Society for Elementary Education in Paris, which Jacques-Joseph had joined. Now that Louis XVIII's government wanted to expand education throughout France, the two brothers were encouraged to set up a Lancaster method school at Figeac.

The Lancaster method of teaching had been developed by Joseph Lancaster, who had opened a school in London in 1798. The method, which soon became popular in Europe, relied on a teacher instructing the older and the more able pupils, who were then set to teaching the rest. With backing from the national government and the local authorities, Jacques-Joseph and Champollion began preparations to set up such a school in Figeac, and so, barely a year after he had written that he was determined to leave public education, Champollion was heavily involved with a new school and a new teaching method. The Ultras and the Catholic Church were both bitterly opposed to new schools, and the Lancaster method in particular, since they were designed to educate the children of ordinary people. Their view was that education should be reserved for the nobility and clergy – it was wasted on the majority of the population and only made the people more difficult to indoctrinate and control. Political circumstances ensured that once again Champollion was inadvertently making enemies.

It was while they were occupied with preparations for the new school that the two brothers were in turn released from exile, but neither of them immediately left Figeac. Only in April 1817 did Jacques-Joseph finally travel to Paris, where he soon became the secretary of Bon-Joseph Dacier, himself the Perpetual Secretary of the Academy of Inscriptions and Literature at the Institute of France. Champollion was left to complete the preparations for the new school, and he soon began to suffer from the stress of his situation at home. In order to protect the future of their sisters, he and his brother had renounced any claim on the family business, including their rights of inheritance, but the bookselling business had continued to decline because of the alcoholism and mounting debts of their father. Unable to cope, Champollion wrote to Jacques-Joseph less than a month after his departure with the reproach that he had left without sorting out the family's problems, of which the most urgent was the need to take legal measures to restrain their father. The next week he reported to his brother that he had managed to secure loans to settle the most pressing debts and prevent the furniture being sold, even accepting help from his friend Thévenet in Grenoble, but was still having problems with his father, who angered him greatly.

Supported by the local authorities and inhabitants of Figeac, Champollion managed to complete the preparations for the new school, which opened in July 1817 with a teacher from Paris and soon had forty pupils, but the strain of the preceding months took its toll on his health, so that he lost weight and suffered from long coughing fits and attacks of fever. He had not done any work on the decipherment of hieroglyphs for over seventeen months now and was not in the best state to start work again, but two events encouraged him to resume his research and to ask Thévenet to send some of his many notes and materials which he had hidden in Grenoble when he was exiled – these events were the arrival of a review of his published book and a meeting with a rival decipherer.

Although it had been published in the *Monthly Review* in London more than a year earlier, Champollion had only just received a copy of an anonymous, lengthy and broadly favourable review of his *L'Égypte sous les pharaons*. Although critical of some of its details, as authors always are, he was nevertheless delighted. The reviewer was in fact Thomas Young (writing anonymously as ever), who remarked that Egypt had always held a fascination for the French and that:

> Perhaps it would be well for Europe if the French were suffered to acquire the country: since the destructive character of the climate would render it an efficacious drain for the superfluous young men of France, who otherwise become troublesome neighbours; and the sympathetic licentiousness of French and Ægyptian manners would facilitate an amalgamation of the people, that might be favourable to the recivilization of an important corner of Africa.

After this xenophobic outburst, Young indulged in a long exposition of Egypt as perceived by him (which demonstrated the narrowness of his view), before finally presenting the favourable review of Champollion's work.

The other event that reawakened Champollion's interest in hieroglyphs was a visit in August from a would-be rival: a Doctor Roulhac from the nearby town of Aurillac. While in exile in Figeac Champollion was outside most academic circles and had even less idea than before of the progress being made by others towards decipherment. Without direct and friendly contact, it was very difficult to find out what other scholars were doing, and when scholars did correspond with one another on friendly terms, they seldom revealed their full results before publication. Even such published results could be difficult to obtain, a problem highlighted in a letter to Young a few years later from the antiquarian and traveller Sir William Gell. Writing from Rome, Gell complained that he could not obtain an *Encyclopaedia Britannica* publication of Young's: 'Whether your book, or pamphlet, or dissertation on Egyptian hieroglyphics be published, or whether it be only presented to your particular friends, I have never been able to discover; but, after repeated trials *in London*, I could not procure it through my bookseller,' and even though he knew Young had presented a copy to the Library of the Vatican, 'whether it be there or not, a public library is always so difficult to get at, and so very useless to the public.' If wealthy English antiquarians could not obtain English publications in London, Champollion languishing in Figeac in rural France had no chance.

During his visit, Doctor Roulhac showed Champollion the 'general etymological system' he had devised. He claimed the system held the clue to deciphering hieroglyphs and proposed a collaboration between them both, but it was obvious that the doctor's ideas were seriously flawed and almost laughable. Although unproductive, the meeting brought home to Champollion the stark fact that in the short period in which circumstances had prevented him continuing his work on the hieroglyphs the situation had changed. If, instead of just a handful of scholars, the problem of decipherment was now being tackled by a country doctor in a nearby rural town, who else was working on it and what were his main rivals doing? The thought that he might not be the first to decipher the hieroglyphs haunted him and drove him to resume his studies. Indeed, unknown to Champollion, and just two months before the visit from Doctor Roulhac, a meeting of the Commission in charge of publishing the *Description de l'Égypte* had been held in Paris at which the antiquarian Louis Ripault announced that he had found the key to the hieroglyphs. One of the savants who had travelled to Egypt with Napoleon's expedition, Ripault had been the librarian of the Institute of Egypt in Cairo. He gave lectures to other learned societies at Paris as well, but his colleagues rejected his ideas, as did

Champollion when he eventually heard about them, yet until his early death at the age of forty-seven in July 1823 (starving himself in order, so he thought, to improve his intelligence), Ripault was to remain a faithful ally of Champollion.

Theories and methodologies now abounded, but none was even one step closer to success, although they were a constant source of anxiety to Champollion. Barely had he received all his manuscripts from Thévenet when he began to question his position in Figeac. Now that Jacques-Joseph had gone to Paris because of the disgust he felt at the way they had been treated in Grenoble, Champollion realized how isolated he was, and yet his memories of Paris were not good, and he was not tempted to follow his brother there. His only other realistic option was to return to Grenoble, where on the surface at least there was now little unrest, but without a job he would be worse off there than in Figeac. He then heard of the possible reopening of the Faculty of Philosophy at the university in which he might be given a teaching post. Although clutching at straws, it was enough to persuade him he should go to Grenoble, and he arrived there with his nephew Ali in October 1817 after an absence of nineteen months.

Champollion was welcomed by the new Prefect, Choppin d'Arnouville, who had been specially appointed to reconcile the population of the region with the government of King Louis XVIII and put an end to the unrest. The university authorities confirmed the plans to reopen the Faculty of Philosophy and hinted that he could be considered for both the chair of Ancient History and the chair of Hebrew. Greatly encouraged by this prospect, he was in the meantime persuaded by the local authorities to set up a Lancaster method school at Grenoble to act as a model school for the region. Once it was known that he had accepted the task, he had the immediate support of the liberals in Grenoble, but inevitably he also became a target for the Ultras and the clergy in their bitter opposition to such schools. Nevertheless, he pressed on with the work, even training a teacher in the Lancaster method himself rather than having one sent from Paris. In early February 1818 the school opened with 175 pupils, and Champollion in one of his ironically self-mocking moods recorded, 'The Prefect is delighted, the Ultras are angry, and I see I am on the right track.' Increasingly involved in the education of young children, with whom he had the easy affinity of the natural teacher, he also became engaged in setting up a Latin school at Grenoble to provide a Classical education.

During the spring the Prefect made Champollion a diplomatic adviser charged to research and sort out the documents relating to the territories ceded by France to the King of Sardinia-Piedmont in northern Italy, which involved liaising with the King's representative, Count Costa. The two men became friends, and a few months later the Count offered Champollion the Chair of History and Ancient Languages at Turin University, a prestigious post with a good salary. Feeling too tied to the Dauphiné area of Grenoble, too committed to so many educational activities there and to his brother's family, and still hoping that the Faculty of Philosophy would reopen and he would be given a post, he reluctantly and most regrettably refused the offer, still loyal to France after all his troubles: 'To leave France is a true emigration for the profit of foreigners, and I like neither foreigners nor emigrés.'

Only now was Champollion at last able to snatch some time to return to the study of hieroglyphs, after a gap of nearly two years caused by the terrible political upheavals. Concentrating initially on the Rosetta Stone, he once again became anxious for a good copy of the inscriptions, but only managed to obtain a carelessly traced copy of the engraving that had by now been published by the Society of Antiquaries of London. He wrote to Jacques-Joseph in Paris in mid-April of his latest ideas, feeling that he would soon complete his work if he could obtain the Egypt expedition's engraving of the Rosetta Stone:

It is beyond doubt that with the engraving of the Commission, I would end up by placing beneath each hieroglyph the corresponding French word and even the cursive Egyptian: as for the Greek it goes without saying. I don't commit myself too much in saying this because my work is three-quarters finished. I know where the hieroglyphic inscription begins and finishes by its relationship to the cursive [demotic] and the Greek. I will prove that *at least two thirds of it are missing*...In my work there is no charlatanism or mysticism. Everything is the result of comparison, and not of a system concocted in advance; I have already found the *articles*, the *formation of plurals* and some *conjunctions*, but that is not sufficient for me immediately to determine the system of this writing. The results of my work have already overturned all the ideas that I had developed on hieroglyphs.

In this same letter to his brother, Champollion referred to the two people whom he considered as his rivals but whom he did not actually fear – Ripault and Jomard. At this time Jacques-Joseph urged his brother to publish his revised Coptic dictionary and grammar, but this was not an easy task without sufficient means of financing the project, and so Champollion insisted that he should forget the idea and that if he himself had any money he would devote it to publishing his work on hieroglyphs: 'I don't have any intention of amassing debts in the abominable times in which we are living.'

To his delight, Champollion received greatly improved copies of the Rosetta Stone inscriptions in mid-June, but by then he was unable to pay much attention to them, being too involved in the school system in Grenoble and especially with his Latin school which was opening in July. He did manage to start compiling a dictionary of hieroglyphs, which Jacques-Joseph also urged him to publish, and on 19 August 1818 he gave a lecture on his latest work to the Académie delphinale, but before too long he was obliged to put his research to one side to concentrate on his schools. Also, despite its low salary, he reluctantly accepted the Chair of History at the Royal College – the institution that had formerly been the *lycée*, and where he had been such an unhappy pupil that he had begged Jacques-Joseph to release him from what he called his 'prison.' In November Champollion, having given up hope of ever being reinstated at the university, began teaching at the same college where eleven years before he had been so relieved to be leaving that he had fainted.

At the end of December he finally married Rosine Blanc, whom he had met in 1813 when she was sixteen years old and he was twenty-two. During their exile in Figeac, both brothers had lost their jobs at the library at Grenoble, which Rosine's father had used as an excuse to break off the engagement and increase his opposition to the match. Champollion's feelings towards Rosine then seem to have cooled a little, and he may even have tried to persuade her against the marriage, but they continued to write to each other, and there is no doubt that she was in love with him. Now her father had relented, persuaded by the Prefect's friendship towards Champollion, and the wedding took place in the modest cathedral of Grenoble – with Jacques-Joseph still very much opposed and absent from the ceremony.

At least superficially, the marriage to Rosine was a happy one, and it is only in a letter to his friend Angelica Palli, the Italian poetess, eight years later, that there is any hint of dissatisfaction. Champollion appears to give a frank summary of his relationship with Rosine, but since the letter as a whole is an expression of his regard for Angelica, with whom he had fallen in love, the absolute accuracy of what he says is open to question. He describes how Rosine continued to love him during their engagement despite 'my marked cooling,' and then says:

I hoped that my absence would change the ideas and intentions of Anaïs [his nickname for Rosine] concerning me, and that she would give up a plan of marriage that nothing made obligatory and which didn't promise happiness to either of us. I was then persecuted: she found in my unhappiness a generous means of persisting in her previous determination. Several suitors, placed in a more advantageous position than mine in the present and for the future, asked for her hand with insistence. Against the wishes of her family, Anaïs refused them; her father, a violent and harsh man, irritated by such an opposition, tormented her each day with his reproaches and overwhelmed her with marks of his displeasure; he deprived her of nearly all her liberty. Finally, my exile came to an end; Anaïs suffered, she was unhappy because of me. Could I choose to do otherwise? My duty was clear. An indissoluble bond unites us. She has found with me the peace and tranquillity which no longer existed for her in the house of her father.

Without any record of Rosine's view of the marriage, it appears that, like most women in France at that time, she had little expectation of emotional fulfilment within marriage and was more than happy with her life with Champollion – if it were not for his letters to Angelica Palli, there would be no evidence that he himself took a different view.

It was not only Champollion who found it difficult to remain steadfast in the quest to decipher hieroglyphs through all the pressures of his other work and the need to earn a living at this time. Thomas Young had been encountering a similar problem, though lack of funds was certainly not an issue for him. Needing to be fitted in with his duties as physician, his own scientific and literary research was to him of 'no immediate concern, except as furnishing me with employment which, from habit, is generally to me more an amusement than a fatigue.' While Champollion was in exile at Figeac, unable to pursue his hieroglyphic research, Young had found himself increasingly involved in other occupations. In 1814 he had become a member of a committee to investigate the degree of danger posed by the introduction of gas into London, especially from the erection of large gasometers in populous areas. Two years later he was a member of a new committee considering weights and measures throughout Great Britain, and about this time he was also asked to provide a report to the Admiralty on shipbuilding. From the end of 1818, he became superintendent of the *Nautical Almanac* and secretary of the Board of Longitude, which instituted a series of rewards for the captain and the crew of the first ship to complete the North-West Passage from the Atlantic to the Pacific Oceans.

A year earlier Young had written to his friend Hudson Gurney about the problems he was encountering:

> With the hieroglyphics I have done little or nothing since I saw you – but I could never get to *the end* of them, as long as any materials exist unexplained and uncompared: I suppose they might furnish employment for an academy of forty members for half a century, and it will be enough for me to have discovered a mine by which others may enrich themselves. But I do mean to try to make out more – and in a year or two I shall publish what I have done – still anonymously as far as the form goes.

Young did succeed in some of his research on hieroglyphs, at the time when Champollion was occupied in setting up educational institutions in Grenoble after his return from exile. By spring 1818 Young had amassed a great deal of text for a pioneering and extensive article on Egypt, and in the summer plates were engraved showing nearly 200 examples of hieroglyphical vocabulary (about forty being more or less correct) deduced from the Rosetta Stone and other monuments and manuscripts, along with further work on the hieroglyphic numerals. He had a few copies distributed to friends that summer, and the whole lot was published anonymously in December 1819 as a supplement to the *Encyclopaedia Britannica*.

Young's article on Egypt began with a description of recent travellers to the country, right up to 1818, followed by a section on what was known of the ancient deities and their mythology. He next discussed the history and chronology, but although he stated that: 'The early history of Egypt claims a much higher antiquity than that of almost any other nation, and is consequently involved in darkness more impenetrable,' he certainly did not put forward any views which conflicted with the accepted Biblical date for the creation of the world. Subsequent topics were the calendar, customs and ceremonies, and the Rosetta Stone inscription, for which he also studied other manuscripts, especially those published in the *Description de l'Égypte*. He deduced that there were three consecutive ancient Egyptian scripts: 'A few specimens from different manuscripts will be sufficient to show the forms through which the original representation has passed, in its degradation from the *sacred* characters, through the *hieratic*, into the *epistolographic*, or common running hand of the country.' By sacred characters he meant hieroglyphs, but by hieratic he actually meant what are today called 'linear hieroglyphs.' These are just monochrome hieroglyphs drawn or painted in outline form, mainly on papyri and coffins, rather than the more elaborate, detailed and usually coloured hieroglyphs found on monuments and in tombs. Confusingly, these linear hieroglyphs are now also sometimes called 'cursive hieroglyphs,' although they are true hieroglyphs and certainly not a true cursive handwriting script like hieratic. Even more confusingly, what is now called hieratic was at times labelled 'epistolographic,' 'enchorial' or 'running hand' by Young, who failed to realize that there were two types of cursive script – hieratic and demotic. Although he noticed that the cursive script became 'indistinct in its forms,' he had not yet worked out that a separate demotic script had evolved: he did recognize, however, that what is now called hieratic (what he refers to in this instance as 'epistolographic') was a script derived from hieroglyphs.

   The *Encyclopaedia Britannica* article was concluded by the 'Rudiments of a Hieroglyphical Vocabulary' (such as the names of kings, animals and numbers), pieces on sounds and phrases, and a description of monuments. In the Rosetta Stone inscriptions, various non-Egyptian proper names could already be read within the ancient Greek text and Åkerblad had earlier identified the same names within the demotic script. For the first time, Young now showed in his vocabulary that one of those names existed within the damaged hieroglyphic text – that of Ptolemy. Ruling Egypt from 204 to 180 B.C., Ptolemy V Epiphanes was the son of Ptolemy IV and the sister he had married called Arsinoë III. A year after his accession, Ptolemy V lost most of Egypt's possessions in Asia Minor, Palestine and the Aegean, and serious revolts erupted in Egypt for the next two decades. His troops were defeated by Antiochus III the Great, who was king of territories around Asia Minor and Syria, and as part of the peace negotiations Ptolemy married Cleopatra, the daughter of Antiochus III – she was the first of a line of Cleopatras that ended with the most famous Cleopatra, who was the last ruler of Egypt before it became part of the Roman Empire.

Despite Barthélemy's and Zoëga's earlier deductions, Young thought he was the first to show that hieroglyphs in rings or cartouches represented names, but he was certainly the first to show that the name Ptolemy occurred six times within the Rosetta Stone hieroglyphic text. Because Ptolemy was a later foreign name (the Greek Ptolemaios) and not an Egyptian one, he was sure that it would be spelled out in hieroglyphs using phonetic alphabetic signs (that is, in hieroglyphs that each represented a single sound), whereas Egyptian names would use hieroglyphs representing ideas: 'The *phonetic characters*, according to the traces which may be discovered in the words Berenice, Ptolemy... will afford something like a hieroglyphic alphabet, which, however, is merely collected as a specimen of the mode of expressing sounds in some particular cases, and not as having been universally employed where sounds were required.' He was correct to the extent that foreign names were spelled out mainly using a limited range of alphabetic hieroglyphs, but he was wrong about Egyptian names because these were written with the whole range of hieroglyphs, not just those hieroglyphs representing ideas. For example, the cartouche ⬭ contains the name of the Pharaoh Menkaure, who built one of the pyramids at Giza, near Cairo, and was later known to the Romans as Mycerinus. As well as ideograms (signs representing ideas) such as ⊔ (*ka*, meaning something like 'spirit' or 'soul'), the cartouche also contains ∿∿∿ (the alphabetic sign for 'n').

On the Rosetta Stone, the name Ptolemy occurs three times in a short cartouche, but three times in a longer cartouche where it is associated with the King's special titles. Young was therefore able to identify, fairly accurately, how the name Ptolemy was formed, even though the original scribe had omitted an essential sign from one of the six cartouches.

Ptolemaios
(Ptolemy)

| Hieroglyph | Young's value | Correct value |
|---|---|---|
| ▢ | p | p |
| ◠ | t | t |
| ∱ | not necessary | o |
| ⋙ | ole or lo | l |
| ⸔ | ma or m | m |
| ⑪ | i | y or ii |
| ⎮ | osh or os | s |

Although vowels are not shown in hieroglyphs, the Egyptians did use some symbols to give an approximate rendering of sounds in foreign names. The hieroglyph used to represent the sound 'o,' for example, has no directly equivalent sound in English or Greek – 'o' is only an approximation.

By chance, Young had also managed to study a copy of an inscription from the huge religious site of Karnak in Upper Egypt which contained the name Ptolemy I Soter. The first in a long line of Macedonian Greek-speaking kings, each called Ptolemy, who ruled over Egypt after Alexander the Great's conquest of the country, Ptolemy I Soter had married Berenice, a noblewoman from Macedonia. In the inscription two cartouches were present, one of which Young recognized as reading Ptolemy. He deduced that the other cartouche would read Berenice, also a foreign name (in Greek, Berenike).

Berenike
(Berenice)

The only shared sign between the Ptolemy and the Berenice cartouches was 𓏭 for 'y,' but Young's interpretation, based partly on his limited knowledge of Coptic, was:

| Hieroglyph | Young's value | Correct value |
|---|---|---|
| 𓃀 | bir | b |
| ⌒ | e | r |
| ⌇ | n | n |
| 𓏭 | i | y or i |
| ▱ | superfluous | k (or hard g) |
| 𓅿 | ke or ken | a |
| ◠ | female termination | female determinative |

Young also believed he had identified other names, including that of the Queen Arsinoë, from inscriptions at the temples of Kom Ombo and Philae and within the zodiac at Dendera. He was still convinced that hieroglyphic alphabetic signs were only used for foreign names and titles, and his failure to recognize that hieroglyphic writing uses various kinds of hieroglyphs for both foreign words and the Egyptian language would prevent him from making further useful advances and lead other researchers astray. Even so, from this and other research he thought that he had worked out fourteen letters of a hieroglyphic alphabet, the first time this had ever been attempted, but in the end only the letters 'f' – ⪗, 'y' – 𓏭, 'm' – ⪦, 'n' – ⸺, 'p' – ▫ and 't' – △ were correct.

One important step was his observation that the hieroglyphic group ⚲ was often added to personal names in papyri and to the end of the name Berenice in his Karnak cartouche and seemed to be a female indicator. This combination of two signs was actually used as a suffix in the names of some goddesses and royal ladies and means 'divine female.'

Young sent a copy of his article to Jomard well before the publication date, who in his reply of September 1819 wrote that he himself had not done any research on hieroglyphs for a long time, and that 'it still has not been possible for me to produce the hieroglyphic vocabulary which I spoke to you about four years ago.' Still regarding himself as a contender in the race to decipher hieroglyphs, Jomard remained unwilling to share information. As for Young's vocabulary, Jomard confessed that he had not found time to study it but it 'must have cost you an infinite amount of trouble. It has only been possible up to now to glance at it.' Young had by far exceeded everything that Champollion had ever published on hieroglyphs (which was virtually nothing), and it seemed that with such a lead he could not be overtaken.

CHAPTER SIX

# (Cleopatra)

It was many months before Champollion saw the published *Ency-clopaedia Britannica* article and realized the progress made by Young, and as he himself had little time to undertake his own research, it looked as if the prize of decipherment was slipping from his grasp. So immersed was he in various aspects of education at Grenoble, encountering constant setbacks in the running of his schools, that he even turned down a renewed offer of a professor-ship at Turin University. Since losing their jobs at the library during their exile at Figeac, the two brothers had lobbied the authorities to be reinstated; in September 1819 Jacques-Joseph was successful and so returned from Paris to take up his post of librarian. However, he still needed to be in the capital to continue the projects in which he had become involved, such as his publication on their findings at Uxellodunum, and so Champollion sacrificed his teaching at the Latin school in order to do Jacques-Joseph's work at the library. On top of his position as Professor of History at the Royal College, this proved a burden because the authorities required a complete cata-logue of the contents of the library. Champollion's health was failing and he was now so exhausted that he declared the 'wish to abandon myself to the sweetness of doing nothing and thinking nothing.'

Just as it seemed that his situation could not be worse, politics once more began to have an adverse effect on his life. Through the intrigues of the Ultras, Choppin d'Arnouville, the moderate Prefect with whom Champollion had worked so closely, was replaced by the Baron d'Haussez in February 1820. As the Baron was one of the Ultras, the balance of power in the region around Grenoble immediately tilted in their favour. When what was happening became apparent, the growth in the power of the Ultras was immediately matched by rising resistance, and as more liberal university staff were replaced by men favoured by the Ultras, students in the law school rose in revolt in May. This unrest was exacerbated a few days later by a visit from the Duke of Angoulême, who was the son of the Count of Artois, the head of the Ultras. The Duke was met with noisy demonstrations, protests and shouted threats, violence broke out in some areas and the visit was curtailed.

As a result of the unrest in the town, instead of ruling with leniency as the previous Prefect had done, the Baron d'Haussez imposed repressive measures and continued to favour the Ultras. Soon Champollion realized that his enemies were intriguing to have Jacques-Joseph dismissed once again from the library, and so in July he wrote to his brother urging him to return from Paris before his absence from Grenoble could be used as grounds for dismissal. Before the letter reached Paris, however, Jacques-Joseph's dismissal was publicly announced in the newspapers. By this time Champollion was in very poor health, suffering from insomnia, stomach pains and fainting fits. His doctor ordered complete rest and he kept to his lodgings for a time, which only aroused the suspicion of the authorities who thought he was plotting something and so put him under continuous surveillance. When he had recovered sufficiently to confront the Prefect, he insisted on his right to remain assistant librarian. Up to this point, he had tried to remain on good terms with the new Prefect, but this proved to be a stormy meeting which brought into the open the Prefect's hostility towards the two brothers. Nevertheless, by October 1820 Champollion had prevailed and even took over the post of full librarian from which Jacques-Joseph had been dismissed.

More than two years had elapsed since Champollion had presented his latest theories on hieroglyphic writing to the Academy at Grenoble. In Paris Jacques-Joseph was by now aware of the progress that Young had been making and informed his brother of this work, continually urging him to publish his own results, which he was still reluctant to do, much as Young was reluctant to publish under his own name. Rather hastily, Champollion was dismissive of Young's publication, even though he had not seen it:

> The discoveries of Dr. Young, announced with so much splendour, are only a ridiculous bragging. The much praised discovery of the claimed key fills me with pity. In all honesty, I feel sorry for the unhappy English travellers in Egypt, obliged to translate the inscriptions of Thebes, the *passe-partout* of Dr. Young in their hands...I beg you to buy it for me straight away from London.

Now he knew he must stake his own claim, and he finally decided to write and publish a booklet to present his ideas on hieroglyphs and hieratic.

Through the winter of 1820–21, various ailments continued to trouble him, and early in 1821 he learned that his father had died at the end of January. Champollion complained of his health, feeling he was no longer good for anything as he struggled to produce his booklet, and the political health of the region was in an even worse state. The harsh measures of the Prefect and the intrigues of the Ultras only produced widespread resentment among the population as more and more liberals were forced from their positions. On 3 March 1821 Champollion was removed from his professorship at the Royal College 'on a temporary basis,' and he was powerless to do anything except use the free time for work towards decipherment. He wrote to his brother, 'It is my Egyptian studies that will win,' and concentrated even harder on his publication.

The discontent within Grenoble came to a head on 20 March,

with a general uprising against the Prefect and his Ultra supporters. Political factions once opposed to each other now found a common cause in demanding a 'free constitution' – one that was free from the attempts of the Ultras to reintroduce all the evils that had prevailed before the Revolution. Very quickly the inhabitants took control of the town and confronted the Prefect, who tried to calm them, and everywhere the white flag of the hated Bourbon kings was replaced by the Revolutionary tricolour. The town gates were shut, all the bells sounded the alarm, and in the Rabot fort across the river from the main town (which protected the only route from Grenoble to Lyons) the garrison was marched out of barracks in case it was needed to put down a rebellion, leaving only a minimal guard. Champollion did not hesitate to lead a small group of men who charged over the bridge and up the steep climb to the fort. Meeting less than token resistance, they dared to raise the tricolour over the fort in place of the Bourbon flag – a symbol of the uprising that could be seen from most parts of the town. When Jacques-Joseph found out what had happened, he was terrified at his brother's boldness, but Champollion joked, 'Perhaps one day the capture of the citadel of Grenoble by an archaeologist, and without bloodshed, will be a point in my favour in the "service records" of the literary profession in these extraordinary times.'

The uprising lasted less than a day, and the troops from the garrison dispersed the angry crowds and restored order with little violence. In the days that followed, thousands of soldiers were camped in the streets and on the ramparts of Grenoble to prevent further protests, and the Prefect, Baron d'Haussez, was intent on reprisals. The leaders of the uprising fled and Champollion lost his post as librarian. Although there was virtually no evidence against him, the Prefect was determined to try him for treason by martial law and denounced him to the government in Paris as a dangerous agitator. Jacques-Joseph hurried back from Paris to see what could be done, while in Grenoble the lawyers were still arguing over whether the uprising was an act of treason against the King or a legitimate protest to bring the distress of the people to his attention. Despite his deteriorating health and despite awaiting trial for treason, Champollion completed the seven pages of text and the seven plates of his booklet in June: 'My last picture with 700 hieroglyphic and hieratic signs has killed me.' Published at Grenoble, his *De l'écriture hiératique des anciens Égyptiens* (On the Hieratic Writing of the Ancient Egyptians) became as difficult to obtain as Young's anonymous publications. The booklet announced that hieratic was a simple modification of hieroglyphic writing, only differing by the form of its signs, a deduction that Young had already made several years earlier, albeit published anonymously in an obscure journal:

> Whether he [Champollion] made this discovery before I had printed my letters in the *Museum Criticum*, I have no means of ascertaining: I have never asked him the question, nor is it of much consequence, either to the world at large or to ourselves. It may not be strictly just, to say, that a man has no right to claim any discovery as his own, till he has printed and published it: but the rule is at least a very useful one.

Champollion's work was independent confirmation rather than plagiarism. He now correctly recognized that hieratic was a form of hieroglyphic handwriting, calling it a 'hieroglyphic tachygraph,' but he made one disastrous error: he regarded all the signs as ideograms and not one of them as phonetic – 'the hieratic characters are signs of things, and not signs of sounds.' It looked as if Champollion was losing his way.

He and many others in Grenoble were saved from punishment for treason by the arrival of the Duke of Bellune, sent by the King to make sure that the investigation into the uprising was conducted with fairness and leniency. The Duke decided that there was no evidence against Champollion to warrant severe charges or a major trial, and instead he was tried before an ordinary civil court, where in early July he was acquitted of all the charges against him, including that of treason. In the end, the act of raising the tricolour flag over the fort hurt him less politically than it did physically. The rapid scramble up the steep hillside had put too much strain on his fragile health, and he was still suffering breathlessness, fainting spells and dizziness. Even if he had been well enough to work, he now had no place at Grenoble, because before the trial his enemies had managed to oust him from all his positions, and his 'temporary' removal from his professorship at the Royal College had been made permanent. On 8 July he wrote to Jacques-Joseph that he had drained the dregs of every bitterness that Grenoble could force on him, no injustice could any longer affect him, and that he had nothing more to lose there: 'The whole universe cries out to me: "Leave! Travel – get away!" Now I am going.' Three days later, his spirit almost as broken as his body, he left Grenoble, once again accompanied by his nephew Ali, on the long journey to Paris.

While Champollion was still awaiting trial in Grenoble in June 1821, Thomas Young and his wife set off on a leisurely tour of Europe. Stopping off at Paris, Young attended a meeting of the Academy of Sciences at the Institute of France and met some of its most illustrious scientists, including the naturalist and geographer Alexander von Humboldt, the astronomer and physicist François Arago, the mathematician and astronomer Pierre Laplace, the astronomer and physicist Jean-Baptiste Biot, and the naturalist Georges Cuvier – some of whom were soon to become good friends of Champollion. Moving eastwards across France through Lyons and over the Alps to Turin, at about the same time as a weak and exhausted Champollion was abandoning Grenoble on his journey westwards via Lyons to Paris, Young and his wife recorded their admiration at the scenery as they travelled south through Italy to Rome and then on to Naples. From there they went to Siena, Pisa and Livorno (Leghorn), where in September Young managed to see the collection of Egyptian antiquities belonging to Bernardino Drovetti, the French Consul in Egypt, which had been amassed by him over many years but had only recently been brought to Italy. It was the first large collection of Egyptian antiquities to arrive in Europe, and it caused a sensation.

Young noticed one object, not entered in the listings of the collection, that was a bilingual (Greek and Egyptian) stone from Memphis, near Cairo, with Greek, demotic and hieroglyphic inscriptions, all barely legible – he immediately thought he had stumbled on another Rosetta Stone. Writing to his friend Hudson Gurney, he informed him of his remarkable find:

Pisa amply repaid us for taking this circuitous route; Leghorn, if possible, still more. But what you will be better pleased to hear, is the discovery that I made of a *bilinguar stone among Drovetti's things*, which promises to be an invaluable supplement to the Rosetta inscription, as I dare say Drovetti is well aware. There are very few distinct hieroglyphic characters about the tablet, and the rings [cartouches] for the name of the king are left blank...Under the tablet are about fifteen lines of the enchorial character, and about thirty-two in Greek.

Young was irritated to be unsuccessful in obtaining casts of the stone:

I engaged a distinguished artist of Florence to undertake the performance of my plan; but I believe he was accidentally prevented from fulfilling his engagement. It appears, however, that his labour, as far as I was concerned, would have been wholly lost; for Mr. Drovetti's cupidity seems to have been roused by the discovery of an unknown treasure, and he has given me to understand, that nothing should induce him to separate it from the remainder of his extensive and truly valuable collection, of which he thinks it so well calculated to enhance the price; and he refuses to allow any kind of copy of it to be taken.

Because he felt the inscriptions on this stone were essential for decipherment, he tried everything he could to induce Drovetti to provide him with a copy, but in vain. Disappointed, Young and his wife next went to Florence, where they found mail from England awaiting them, including a letter with the news that Mrs. Young's mother was seriously ill. Cutting short their planned grand tour of Europe, the Youngs hurried back through Switzerland and along the Rhine Valley, but they had only reached Geneva when they were informed that Mrs. Young's mother was dead. It was October when they both finally arrived back in England.

Nearly twelve years had passed since Champollion had last been in Paris, but little seemed to have improved when he arrived there from Grenoble on 20 July 1821, just missing the Youngs who had recently left Paris en route for Italy. The politics may have changed, and there was no longer tumult in the streets acclaiming imperial victories or rioting for lack of food, but the city under the new monarchy was shabbier and more squalid than ever and its population had greatly increased. In a state of extremely poor health, Champollion endured the gruelling journey from his adored Grenoble, only reconciled to living in Paris by his desolation at the way he had been hounded from his own town. Illness, despair and exhaustion had brought his life to its lowest ebb and he was sustained only by what his enemies had not managed to take from him: his research into ancient Egypt and the support of his family.

Jacques-Joseph was lodging with a friend in the rue des Saints-Pères, and space was found there for his brother and Ali as well, but by the end of the summer they moved to a larger rented house at 28 rue Mazarine, where there was room for Champollion's wife Rosine, who travelled from Grenoble to join him, although Jacques-Joseph's wife Zoé remained in Grenoble with her family. Champollion took over the attic room as his study – it had previously been used as a studio by the famous artist Horace Vernet, whose paintings of Napoleon's glorious victories on the battlefield came to form part of the Napoleonic legend.

The house in the rue Mazarine was ideally placed for Champollion's research, just yards from the Institute of France with its Academy of Inscriptions and Literature, where Jacques-Joseph still worked as Secretary to Bon-Joseph Dacier, the Perpetual Secretary of the Academy – the same Academy that had rejected the publication of Champollion's Coptic dictionary and grammar six years earlier and had even blackballed Jacques-Joseph's full membership in 1816. The Institute of France, on the south bank of the River Seine, facing the Louvre, comprised four other academies: the Academy of Fine Arts, the Academy of Sciences, the Academy of Political and Moral Sciences, and the exclusive Académie Française to which the first woman was elected in 1980. The rue Mazarine was also fairly near the College of France, where Champollion had been a student, and not far across the river were the Royal (National) Library and the Egypt Commission, where work continued under Edme Jomard on the volumes of the *Description de l'Égypte*.

In the weeks before the move to rue Mazarine, Jacques-Joseph started the physical and social rehabilitation of his brother. As well as suffering from exhaustion and illness, Champollion became easily depressed, sensing that he was just on the brink of deciphering the hieroglyphs, but all too often feeling that death would cheat him of his achievement. It was as much his mental state as his physical state that required nursing, and Jacques-Joseph constantly assured him 'You must live and moreover you will live!' In addition, Jacques-Joseph immediately took steps to reintegrate him into academic circles in Paris and reunited him with Joseph Fourier. After being appointed Prefect of the Rhône in 1815, Fourier had soon been dismissed when the monarchy was restored, and subsequently eked out an impoverished existence in Paris while continuing his scientific work. Champollion was also introduced to Dacier, who showed keen interest in his work and became a powerful supporter. Other new friends included François Arago, Jean-Baptiste Biot and Georges Cuvier, who had been introduced to Thomas Young only a few weeks earlier, as well as many in the newly formed Society of Geography that would officially open in November. Even though Fourier's maxim was 'science inspires a universal friendship,' Champollion's friends were still outnumbered by Royalist enemies, as well as by academic rivals like Jomard and influential supporters of Young, who all believed they were still on the right route to deciphering the hieroglyphs.

Both as a boost to his flagging confidence and to raise his profile among the intellectuals of Paris, Champollion was persuaded to present some of his work to the Academy of Inscriptions. At the end of August 1821, barely a month after his arrival in Paris, he read a report about his conclusions on the hieratic script, based on the ideas in the booklet that he had not long completed in Grenoble. He was only too well aware that the report would provide ammunition for his rivals, joking that he was leaving 'his entrenched position to face the fire of the batteries,' but the lecture was well received and proved to be the first step to establishing his credibility in the Academy of Inscriptions. He was accepted into the academic circles of Paris, and Rosine created a welcoming home in the rue Mazarine, where her husband could return the hospitality he received in meetings at the homes of other researchers in a wide range of disciplines – meetings that included enemies and rivals as well as friends and supporters and often ended in extremely lively debates. Still far from fully recovered, but with a growing belief that he was following the right course, Champollion began an intensive study of demotic, comparing it with Coptic, in preparation for a comparative study of the demotic, hieratic and hieroglyphic scripts, now convinced that they should be regarded as different from each other. That he was left in peace to do his research was in part because decipherment was not uppermost in the minds of the scholars of Paris at that time – Egyptian zodiacs were instead the cause of much controversial discussion, because they were seen as the new, exciting key to the chronology of ancient Egypt and the date of the creation of the world.

In September, the Dendera zodiac arrived in the port of Marseilles from Egypt and was destined to be the talk of the town on reaching Paris the following January after a period of quarantine. This zodiac was a sculpture portraying a circular diagram of astrological symbols that had formed part of the ceiling of a room in the temple of the goddess Hathor at Dendera, some 300 miles south of Cairo. First recorded by the artist Vivant Denon and the engineers Prosper Jollois and Édouard de Villiers du Terrage during Napoleon's Egypt expedition, it had remained a focus of controversy because of various scholars' attempts to use it to calculate the age of the temple. On hearing about the zodiac, the antiquarian and collector Sébastien Louis Saulnier had commissioned the engineer Jean Baptiste Lelorrain to retrieve it from Egypt and transport it to France, and in January 1821 Lelorrain had obtained a travel permit from Mehemet Ali, the ruler of Egypt, who at that time was allowing wholesale looting of the ancient monuments in Egypt.

The British Consul in Egypt, Henry Salt, was in open competition with the French Consul Bernardino Drovetti, both of whom had a tacit agreement that the French had rights to the monuments on the east bank of the Nile and the British had rights to those on the west bank. As Dendera is on the west bank and so 'belonged' to the British, Lelorrain had to proceed secretly. He arrived at Dendera in March, but finding a group of British travellers there he continued up the river to Luxor. When he returned, he found only Egyptians at Dendera and hurriedly organized a gang of workmen. He saw that the zodiac was on two stone blocks, measuring a total of 'twelve feet long, eight broad, and three thick; its weight therefore could not be much less than twenty tons...But as two feet at each end contained only wavy lines or zig-zags, he determined to cut these off, and with his chisels to reduce the stones to half their thickness.' In three weeks he succeeded in cutting the sculpture from the temple ceiling 'by means of saws and chisels and gunpowder,' and it was loaded aboard his boat, but the captain made excuses to delay the sailing. Lelorrain eventually found out that the American lawyer and diplomat Luther Bradish had stopped at

Dendera and found out what was going on. He had bribed the captain to delay sailing and had probably already reported the matter to Salt's agents farther down the Nile. Lelorrain was forced to match Bradish's bribe to induce the captain to sail and reached Cairo in June, only to find that the British Consul had already protested to Mehemet Ali. With his usual cavalier attitude to the monuments, Mehemet Ali enquired whether Lelorrain possessed a permit, and on being informed that he did, allowed the export of the zodiac to France. The British Consul and his friend William Bankes, who was in the process of taking to England an important obelisk from the island of Philae at Aswan, were particularly annoyed at this decision, having been on the point of removing the zodiac themselves – although Bankes later hypocritically protested, 'I have always deprecated, in the strongest manner, such spoliations of existing and entire monuments.'

Many scholars in France and abroad were appalled at this vandalism, and protested about the action of Lelorrain; Champollion was especially angry that the zodiac had now been separated from the hieroglyphic inscriptions that originally flanked it. He had a letter published in the *Revue encyclopédique* in October 1821, asking why it had been thought necessary to hack the zodiac from its context, when a cast of it could have been taken. Other scholars protested openly, including his rival Jomard, but by this time Champollion had learned to be discreet and his own letter was published anonymously. The sculpture is now on display in the Louvre Museum, and a cast occupies its original position in the temple at Dendera.

Apart from the controversy over its removal from Egypt, the zodiac renewed the debate that had continued for some years over its use as a means of dating the Egyptian civilization. Unable to gain any clues from the hieroglyphic texts, which when deciphered would provide detailed evidence of dates and historical events, the scholars relied on interpretations of zodiacs. Reproductions of zodiacs, including the one at Dendera, were minutely studied because it was thought that the positions of the stars shown in them would be their actual positions at the time when the zodiacs had been originally drawn up. If that was the case, calculating the dates when the stars were in those positions would give the dates of the zodiacs. Disagreements broke out between the scholars about dates calculated in this way, but such arguments were nothing compared to the opposition they generated within the Catholic Church. What was in dispute was nothing less than the date of the creation of the world. Some scholars, most notably Jomard, were bold enough to suggest that the Dendera zodiac was many thousands of years old, perhaps as much as 15,000 years old, a concept in conflict with the accepted Christian belief, using evidence from the Bible, that the world had only been created about 6,000 years ago. Champollion did not share the opinion that the zodiac was of great antiquity as he felt that the style of the sculpture showed that it belonged to the Greek or Roman period in Egypt, and it would not be many months before he could read several of its related hieroglyphs and prove its date, but for now, apart from occasionally voicing his opinions in meetings, he concentrated on his work on the demotic script.

Through the autumn of 1821, as Champollion became more optimistic about his line of research and was totally absorbed in his work, events were taking place in England that were to have a profound effect on his future. *The Gentleman's Magazine* reported that an Egyptian obelisk belonging to Mr. Bankes had just been safely unloaded at Deptford and was awaiting transportation to his home in Dorset. At about the same time Champollion's rival Thomas Young was returning home prematurely from the Continent, his grand tour having been cut short by the death of his mother-in-law. If the obelisk had not been brought to England, it is possible that Champollion's decipherment of the hieroglyphs might have been delayed for a few months, but it is beyond doubt that the bitterness and envy which were later to surround his achievement would have been greatly reduced if Young had not returned when he did. As it was, Young's involvement with Bankes and the obelisk later provided him with a slender excuse to attack Champollion and his work.

Just four years older than Champollion, William John Bankes was the eldest surviving son of the antiquarian Henry Bankes of Kingston Hall (now known as Kingston Lacy House) near Wimborne in Dorset. Henry Bankes had been a trustee of the British Museum and a Member of Parliament for Dorset, and William was to become a fanatical collector of antiquities. Unable to undertake the traditional grand tour because of the Napoleonic Wars, he instead served as a Member of Parliament for Truro in Cornwall from 1810 to 1812. He gave up this potentially promising career in order to travel, equipped with numerous letters of introduction from his friend the poet Lord Byron, whom he had met as a student at Cambridge University – it would be nearly eight years before he set foot in England again.

Initially Bankes went to Spain, where Wellington was fighting the Peninsular War, which was then the main area of conflict against the French. He accompanied Wellington in an unofficial capacity, and then spent some time living with Gypsies in Granada. Next Bankes turned his attention to Egypt and Nubia, and from September 1815 he travelled up the Nile as far as Abu Simbel, where the larger of the two rock-cut temples, the great temple of Ramesses II, had not long been discovered by the Swiss traveller Jean-Louis Burckhardt, although it would be two more years before the whole of the temple with its incredible colossal statues was uncovered, because over two-thirds of it was engulfed by drifting sand, to a height of fifty feet in some places. Abu Simbel was and still is a desolate area, just thirteen miles north of the border with Sudan and 150 miles south-west of the farthest point ever reached by members of Napoleon's expedition.

On his return journey northwards Bankes stayed on the island of Philae, which was dedicated to the goddess Isis, the symbolic mother of the Egyptian pharaohs or kings, to explore the ruins of the temples. He became fascinated by a fallen obelisk and its possible base nearby, whose location had already been published in an engraving in the *Description de l'Égypte*. The following year in 1816 Giovanni

Battista Belzoni, a former circus strongman from Padua in Italy, claimed possession of the obelisk in the name of Britain's Consul Henry Salt, who in turn surrendered it to Bankes. This caused much dispute with the French Consul Bernardino Drovetti, but Belzoni finally managed to haul the obelisk, which is twenty-two feet high and weighs approximately six tons, to the edge of the Nile. Before it could be loaded on a boat, the pier collapsed and the obelisk was nearly lost in the river, but against the odds Belzoni managed to retrieve it, load it on a boat and transport it downstream.

He delivered the obelisk safely to the port at Rosetta, where it remained for a couple of years until the base, which had been left on the banks of the Nile because it was too heavy to be carried with the obelisk, was brought to Rosetta by another of Salt's agents. Loaded on a ship called the *Dispatch*, the obelisk and base arrived off the coast of England in June 1821, but the ship was quarantined because the plague had been raging in Egypt when the ship had left. The obelisk and base were finally unloaded and transported to Kingston Hall, but because of damage incurred during the journey, these monuments were left lying on the lawn in front of the house until 1827, when the Duke of Wellington visited Bankes and was asked to lay a foundation stone. It took until 1839 before the obelisk was erected, and only two years later Bankes, a homosexual, was the centre of a public scandal that forced him into exile abroad rather than face trial.

Nowadays greatly eroded by the English weather, the four faces of the Bankes obelisk once carried clear hieroglyphic inscriptions with two different cartouches, while its base was inscribed with a Greek text which included the names of Ptolemy VIII and his wife Cleopatra III. Hated by the people of Alexandria and nicknamed 'Physkon' (pot-belly), this Ptolemy was king of Egypt jointly with his brother Ptolemy VI from 170 B.C., until agreeing to rule the neighbouring kingdom of Cyrene (now Libya) seven years later. He returned to Egypt in 144 B.C., when he murdered his nephew Ptolemy VII and married his own sister Cleopatra II. Two years later, without divorcing Cleopatra II, he also married Cleopatra III, who was the daughter of his brother Ptolemy VI and of his own sister and wife Cleopatra II who had been previously married to Ptolemy VI.

A lithograph of the Greek and hieroglyphic texts was widely distributed by Bankes, and Young received a copy. It was based on drawings made by artists when the obelisk was first unloaded at Deptford, although the Greek inscriptions on the base were only found during the cleaning of the stone when it arrived at Kingston Hall. As he was himself interested in the decipherment of hieroglyphs, Bankes had already noted that one of the cartouches on the obelisk was similar to that identified by Young as Ptolemy on the Rosetta Stone and concluded that its other cartouche was probably Cleopatra, particularly as the Greek inscription on the base contained both these names, even though it was not a copy of the hieroglyphic text. This important deduction was made before Champollion came to the same conclusion, but Bankes did not publish his findings, merely pencilling the single word 'Cleopatra' in the margin of the lithograph. He communicated his ideas to his friends, including Young, who was unable to make any further progress, which he attributed to a mistake by the artist:

It so happens that in the lithographical sketch of the obelisc of Philae, which had been put into my hands by its adventurous and liberal possessor, the artist has expressed the first letter of the name of Cleopatra by a T instead of a K, and, as I had not leisure at the time to enter into a very minute comparison of the name with other authorities, I suffered myself to be discouraged with respect to the application of my alphabet to its analysis, and contented myself with observing that if the steps of the formation of an alphabet were not exactly such as I had pointed out, they must at least have been very nearly of the same nature.

Champollion was not informed of the discovery of the possible Cleopatra cartouche, either by Young or by Bankes, who consistently refused to help him, and Young believed that until more bilingual inscriptions were found, little further progress could be made.

On his thirty-first birthday, 23 December 1821, the idea came to Champollion that as part of his comparison between the demotic, hieratic and hieroglyphic scripts, he should do a numerical analysis of the texts on the Rosetta Stone. To his surprise, the 486 words of the Greek text were paralleled by 1,419 hieroglyphic signs. He had been working on the theory that the hieroglyphs were primarily ideograms, each representing an idea and therefore a single word, but clearly such a large difference between the number of Greek words and the number of hieroglyphs made that theory impossible. He next tried to identify groups of hieroglyphs, but this produced a number of about 180, which again was too far from the 486 Greek words to provide a match. He could not establish a numerical relationship between the Greek text and the hieroglyphic text, and so the obvious conclusion was that there was variability within the hieroglyphic text – it could not be solely, or even mainly, composed of a single type of sign (pictograms, ideograms or phonetic symbols), but must be some sort of combination of two or more types of sign. Champollion realized that the hieroglyphic text must be at least partly phonetic (representing sounds), and from this time onwards he maintained a much more flexible approach than his rivals, because he was rapidly becoming aware of the complexity of the system of hieroglyphic writing.

By the New Year, Champollion was completely immersed in his comparative analysis of the demotic, hieratic and hieroglyphic scripts. Even though he could not yet read any of these scripts, he constantly tried to transliterate the later Egyptian demotic texts sign by sign into the earlier hieratic, and then transliterate the hieratic into hieroglyphs. The more he worked at this method, the more success he had, and he gradually built up his understanding of the three scripts, how they worked, and how they related to each other.

In essence, Champollion took a holistic approach, and looked at all aspects of the ancient Egyptian system of writing, in marked contrast to Young, who relied heavily on rare bilingual texts in the hope that once one or two hieroglyphs could be understood, all the rest would follow easily. Many years later a British Egyptologist, Sir Peter Le Page Renouf, summed up Young's method: 'He worked mechanically, like the schoolboy who finding in a translation that *Arma virumque* means "Arms and the man," reads *Arma* "arms," *virum* "and," *que* "the man." He is sometimes right, but very much oftener wrong, and no one is able to distinguish between his right and his wrong results until the right *method* has been discovered.' Renouf's example shows the complexities of languages: the Latin word for 'and' was usually *et*, but *que* could be used as a suffix for emphasis – in this case added to *virum*, the word for 'man.' The man was Aeneas, as these are the opening lines of *The Aeneid*, Vergil's epic poem: *'Arma virumque cano'* – 'I tell of arms and the man.' This difference in approach between Champollion and Young was to prove crucial, for while many researchers were using the same technique as Young, few had anything like the knowledge of the scripts and related languages that Champollion had acquired, and nobody else was using his methods.

A year earlier, during the winter of 1820–21, a Mr. Casati (described as 'an Italian speculator' by Young) had been travelling in Egypt and discovered a collection of mainly Greek papyri in a pottery jar at Abydos. When these arrived in Paris, Champollion found one papyrus written in demotic which was very similar in its preamble to the demotic text on the Rosetta Stone. He recognized the name Ptolemy and suspected that another name in a demotic equivalent of a cartouche would be that of the Queen Cleopatra. By applying his technique of comparing scripts, he converted this demotic name into hieratic and then into hieroglyphs, arriving at a hypothetical hieroglyphic version of Cleopatra. All he needed now was to match it with a genuine hieroglyphic version of the name to test if his system worked.

In January 1822 Jean Letronne, a specialist in ancient Greek who had been a student with Champollion in Paris, received a copy of the lithograph of the inscription on the Bankes obelisk and passed it on to Champollion. With growing excitement, Champollion immediately recognized the name Cleopatra written in hieroglyphs, because it was a close match for his hypothetical version derived from the Casati papyrus. As yet, it was only a matter of one hypothesis corroborating another, but he was now convinced that his method was correct and it was only to be a matter of time before he could prove it. By comparing the hieroglyphs for the name Cleopatra on the Bankes obelisk with the hieroglyphs for the name Ptolemy on the obelisk and on the Rosetta Stone, Champollion was encouraged that most of the signs that should be common to both names (the hieroglyphs for 'p,' 'o' and 'l') were in the right place to spell these names alphabetically (as 'Ptolmes' and 'Cleopatra'). He deduced that the values of the individual signs were:

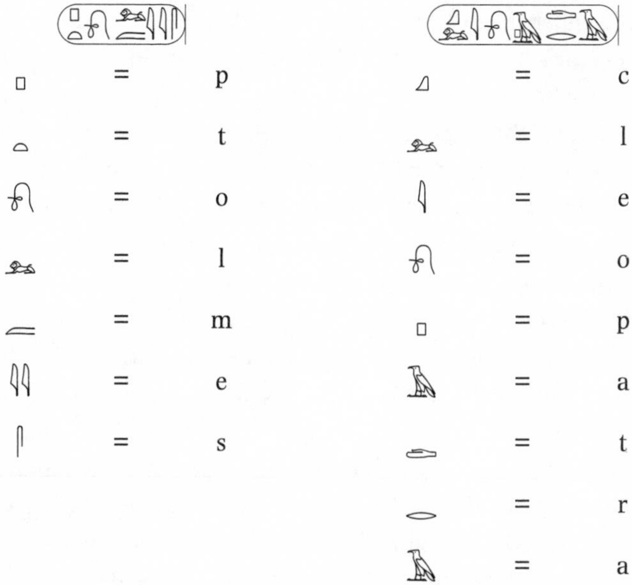

| | | | | | |
|---|---|---|---|---|---|
| ▫ | = | p | ◿ | = | c |
| ◠ | = | t | � | = | l |
| 𓂋 | = | o | ⟨ | = | e |
| ⟷ | = | l | 𓂋 | = | o |
| ⌒ | = | m | ▫ | = | p |
| 𓏺𓏺 | = | e | 𓅓 | = | a |
| 𓈖 | = | s | ⟶ | = | t |
| | | | ◠ | = | r |
| | | | 𓅓 | = | a |

Of the hieroglyphs that should match, only the value for 't' differed, and so he thought that at least two different hieroglyphs ⌒
and ⇐ could be used to express the sound of 't.' He designated
these hieroglyphs homophones (having the same sound), and
rather than being discouraged at the extra complication, he realized that these signs would account for some of the apparent
complexity within the ancient Egyptian scripts. Champollion took
it as a good omen that the recumbent lion ⚊, the hieroglyph used
to represent 'l,' occurred in both Cleopatra and Ptolemy. Referring
to his own use of the lion motif, which he had adopted in childhood,
he prophesied: 'These two lions will help the lion to victory!' When
Champollion's enemies later claimed that he had based his decipherment system on the identification by Bankes of the name
'Cleopatra,' Young made the incredible claim that because he himself had 'inspired' Bankes to identify the name, then Champollion's
decipherment of the hieroglyphs was also due entirely to Young.

Champollion was now certain that in the texts of Ptolemaic
Egypt, alphabetic hieroglyphs were used to spell out non-Egyptian
names (such as Ptolemy and Cleopatra), noting: 'If the Egyptians
wished to show a vowel, a consonant or a syllable in a foreign name,
they used for that a hieroglyphic word expressing or representing
an object whose name, in the spoken language, contained in full or
in its first part the phonetic value of the vowel, of the consonant or
of the syllable that was to be written.' This use of hieroglyphs to represent individual sounds was still thought to be a late development,
not used before the Ptolemaic Greek rule of Egypt from the fourth
century B.C., but it had provided Champollion with his first firm
foothold. Although progress towards decipherment would now
advance by giant strides, he was not in any hurry to make his
results public. Having learned by bitter experience the pitfalls of
premature publication (with the introduction to his *L'Égypte sous
les pharaons* in 1811), he was yet to learn the pitfalls of prevarica-
tion. When in March he published an account of the Bankes obelisk
and its Greek and hieroglyphic inscriptions in the *Revue ency-*

*clopédique*, he only hinted at the progress he was making, without giving any details. Some of his findings have now been refined: 𓏭 is now transliterated as 'ii' or 'y' not 'e'; 𓇋 is transliterated as 'i' not 'e'; 𓐍 is usually transliterated as 'k' or 'q' rather than 'c'; and although homophones do exist, ◠ and ⬤ are not true homophones for 't,' as ⬤ is usually transliterated as 'd.' The Greek names Ptolemaios and Cleopatra were therefore most likely spelled out in hieroglyphs as Ptolmys and Kliopadra, which gives an idea of the original Egyptian pronunciation.

Using the alphabetical hieroglyphs that he had worked out from the names Ptolemy and Cleopatra, Champollion applied his system to other names in texts of Ptolemaic or Roman date, mainly using the drawings of cartouches that had so far been published in the *Description de l'Égypte*. Gradually he was able to deduce the value of other signs and to work out the names in the cartouches of almost all the Greek and many Roman rulers of Egypt, from Alexander the Great, who had taken control of Egypt in 331 B.C., to Antoninus Pius, a Roman Emperor who died in A.D. 161. Champollion also managed to recognize the words

Autocrator     Caesar

'Autocrator' (the Greek word for 'Emperor') and 'Caesar' were titles both used in the Roman period, but he still maintained the view that the phonetic hieroglyphs within these names and titles had only been used after the Greek conquest of Egypt in 331 B.C., and had not been used in that way before. His main problem at this stage was the inaccuracy of many plates published in the *Description de l'Égypte*, a point on which he constantly and tactlessly commented, much to the anger of Jomard.

As well as making this direct progress towards decipherment in January 1822, Champollion could no longer resist becoming involved in the controversy surrounding the Dendera zodiac, whose arrival in Paris that same month was causing such a sensation. It was put on temporary display in the museum at the Louvre, and people who knew little or nothing about zodiacs queued for hours to see it. The whole of Paris 'only thought of that, only saw that, only spoke of that,' not least the scholars who professed an interest in ancient Egypt. Public pressure became so great that King Louis XVIII bought the zodiac for the Royal Library at the enormous price of 150,000 francs, and it remained there until 1919, when it was transferred to the Louvre. Even as public interest waned, the scholars were only just getting into their stride, preparing academic papers on the significance of the zodiac, its date and its importance for ancient Egyptian chronology, and so reopening all the controversy that had previously attracted such hostility from the Catholic Church. Soon the scholars were offering a choice of 'so many opinions, all very learned, but all very different.'

The famous astronomer and physicist Jean-Baptiste Biot, one of the new friends Champollion had made on his return to Paris, prepared a lengthy report on the dating of the zodiac. By supposedly identifying the stars shown on the sculpture and calculating the year in which they could have been seen in those relative positions in the sky, Biot dated the zodiac to 716 B.C. Despite a friendly warning from Champollion that his method was wrong, Biot made his ideas public at a series of meetings of the Academy of Sciences and the Academy of Inscriptions in July, but later that month Champollion completely demolished his theory in a letter published in the *Revue encyclopédique*. Biot had based his conclusions on the fact that the figures and the groups of hieroglyphs on the sculpture often had an associated star, which he assumed had been used to mark the position of a real star in the sky. In a masterpiece of analysis, Champollion showed that Biot's theory was inconsistent because it could not account for all the figures marked by stars – no pattern of major stars matched the pattern on the zodiac. He explained the star symbols as 'signs of the type' (now called 'determinatives'): hieroglyphs that determined the nature of the figures or the nature of the group of hieroglyphs associated with the figures on the sculpture. Rather than marking the position of an actual star, Champollion maintained that the star symbols meant that what was being described in the hieroglyphs was a star or something connected with the idea 'star,' such as a constellation: 'The star of the inscriptions of Dendera is therefore the *last* hieroglyphic sign of each of them, and must be considered, not as the representation of a star, but as a simple element of the hieroglyphic writing; that is to say, as a kind of *letter*, and not as an imitation of an object.'

Although little notice was taken of this deduction at the time, Champollion had made another major step towards decipherment. Young had already noted that a determinative sign meaning 'divine female' usually accompanied the names of goddesses and queens, but Champollion had found another determinative and reported that he had found a few others in the Rosetta Stone inscription – he was soon to find many more. Determinatives form a class of hieroglyphs whose function is to clarify the meaning of other groups of hieroglyphs, so their recognition was a significant advance in the decipherment process. For example, ∆ is a determinative showing that the group of hieroglyphs associated with it conveys an idea of forward movement, often running or walking, and the determinative sign 𝄪 means 'enemy' or 'foreigner' (similar ideas in ancient Egypt) and so indicates that the group of hieroglyphs that precedes it means something to do with an enemy or a foreigner. That the enemy is shown as a man with his hands tied behind his back was due to the Egyptians' religious and magical beliefs. Because it was thought that magic spells could bring to life pictures and sculptures, any potentially dangerous representations had to be neutralized against the possibility of them being given life accidentally – in this case, the enemy is shown as a helpless captive. The determinative sign 𝄪 is sometimes called 'the little bird of evil' because it signifies anything small, weak or bad: concepts closely connected in the ancient Egyptian language.

Some hieroglyphs only function as determinatives, but many other hieroglyphs were sometimes used as determinatives as well as their normal function. The implications of the existence of determinatives were only just beginning to be understood – if a sign was needed to make the meaning of a group of hieroglyphs easy to understand, it implied that a group of hieroglyphs could have more than one meaning, as Champollion was later to discover. Determinatives can actually radically alter the meaning of a group of hieroglyphs. For example, the hieroglyphs ⟨≈○⟩ can be used with the determinative ⊙ to form the group ⟨≈○⊙⟩ meaning 'time,' but used with ⟨⟩, the resulting group of hieroglyphs ⟨≈○⟩ means 'weak' or 'limp.'

In London, Young had done very little productive research into Egyptian writing since the extensive article he had written and published at the end of 1819, but he continued to amass material and formed a small Egyptian Society with the purpose of copying and publishing, though not deciphering, all hieroglyphic inscriptions. He was still trying to obtain a copy of Drovetti's bilingual inscription, writing to Gell in May 1822 that: 'I am at present resolved to wait for Drovetti's inscription, which I saw at Leghorn, before I publish the details of my translation of the Rosetta Stone.' In Paris, Champollion's work on comparing the different Egyptian scripts was continuing: his study of the demotic script was nearing completion, and he was invited to give various reports on hieratic and demotic to the Academy of Inscriptions in July, August and September 1822. He now fully grasped the true relationship between hieroglyphs, hieratic and demotic, that hieratic derived from hieroglyphs and demotic derived from hieratic, and that all of them corresponded to the same language (albeit a language which had substantially changed over time) and had more or less the same rules: once one script was deciphered, they could all be deciphered.

Although the audiences contained many old rivals and enemies, his reports to the Academy were enthusiastically received, and at one meeting, to Champollion's great surprise and immense pleasure, Silvestre de Sacy stood up and warmly praised his work. Champollion had idolized de Sacy when he had been a student in Paris and had been deeply wounded by his former professor's subsequent hostility, based more on his political prejudice as a Royalist than on his opinion of Champollion's ability. On being reconciled with de Sacy, he was filled with happiness and optimism that at last he had been accepted by the academic establishment.

For more than a year, Champollion had been concentrating solely on the problem of decipherment, although before that he had been working on it intermittently for more than twenty years, tragically too often interrupted for financial and political reasons. He could now easily transliterate demotic to hieratic as well as hieratic to hieroglyphs, even though these three scripts could not themselves be read. He was confident that he had identified many of the hieroglyphs used phonetically to spell foreign names and titles when Egypt was under Greek and subsequently Roman rule, and he could therefore read the names of most Greek and Roman rulers of Egypt. From his analysis in December the previous year of the number of hieroglyphs representing the Greek text on the Rosetta Stone, he knew that not all hieroglyphs could be ideograms and was becoming aware that the use of determinatives implied the existence of groups of hieroglyphs that looked the same but had different meanings. As a fluent Coptic speaker, he was well able to work out the possible meanings of words he was just beginning to decipher, because Coptic words were often similar to those spoken some 2,000 years earlier: for example, the Coptic for Egypt is *keme*, while the ancient Egyptian word was *kmt* (pronounced something like *kemet*), and the Coptic word for good is *nufe*, while the ancient Egyptian word was *nfr* (usually pronounced as *nefer*). After the success of his report in August to the Academy of Inscriptions, Champollion was working in a state of excitement and heightened awareness. The relevant material was at his fingertips, and decipherment was so close it was almost tangible. He eagerly seized on each new copy of hieroglyphic text that came his way, searching for the connections that would give him the keys to how the hieroglyphs worked – keys not just to the writing system, but to history itself.

Rising early on the morning of 14 September 1822, anxious to resume his research, Champollion received in the post copies of drawings of hieroglyphs on the temple of Abu Simbel, done by the distinguished architect Jean-Nicolas Huyot, who had recently travelled in Egypt and Nubia and was renowned for the accuracy of his drawings and the reliability of his notes. When Bankes had visited the temple of Abu Simbel just a few years earlier, it was barely visible, but in 1816 the entrepreneur Belzoni spent weeks clearing tons of sand from the façade, only to run out of money before the work was completed. Returning the following year, he spent three further weeks clearing sand from the central doorway, finally entering the temple on 1 August 1817 to discover an immense room with incredible decoration and hieroglyphs. Sadly, the original discoverer of the temple, Jean-Louis Burckhardt, died in Cairo of dysentery at the age of just thirty-two, before the news of Belzoni's find reached him.

Poring over the drawings in his attic room in the rue Mazarine, Champollion soon noticed names within cartouches – names that he had never seen before. The first sheet contained , and he immediately recognized its first sign ⊙ as a picture of the sun. He knew that in Coptic the word for sun was Re or Ra, which also happened to be the name for the ancient Egyptian sun god. From his earlier work he knew that the last two signs  would transliterate as 's' in Ptolemaic or Roman names, which if applied to this cartouche would give 'Ra ... ss,' or more likely 'Ra ... ses' because vowels were not normally shown in hieroglyphs. At once he saw that if the other sign  was 'm,' it would represent 'Rameses,' a name known to have been used by several pharaohs well before the Greek and Roman rule of Egypt – which is nowadays spelled as Ramses, Rameses or Ramesses. With mounting excitement and joy as he began to understand what was happening, yet still fearful that he would find proof that his system was totally wrong, he searched the rest of the Abu Simbel drawings and found the name ⟨𓅓𓏠𓋴⟩. Once again he read 𓏠𓋴 as 'mes,' and he recognized the sign in front as a picture of an ibis, recorded by ancient writers as the symbol of the god Thoth who was revered by the Egyptians as the inventor of hieroglyphs and god of scribes. The name in the cartouche therefore read 'Thothmes,' better known nowadays by the ancient Greek version of the name, Tuthmosis – another name used by several pharaohs well before Greek and Roman times. In his *Encyclopaedia Britannica* article of 1819, Young had also recognized the symbol of Thoth in a cartouche and had deduced but not actually proved the name Tuthmosis. Unlike Young, Champollion instantly saw the underlying principle, and it confirmed the system of decipherment that he had been painstakingly putting together, piece by piece, over the last few months.

He checked and rechecked his discovery until he was certain he was right. Elated, he felt he must tell someone about this breakthrough – he must tell his brother. Gathering up an armful of his papers, he rushed down the stairs from his attic room and into the street, heading for the Institute of France with its imposing dome looming on the skyline a short distance away. By the time he found Jacques-Joseph he was out of breath, and in his excitement was shouting '*Je tiens l'affaire!*' ('I've found it!'), but he had hardly begun to describe just what he had found when he fell to the floor, apparently dead.

# CHAPTER SEVEN

# (An Acquaintance of the King)

Legend has it that after his collapse Champollion was taken home to bed where he remained for a full five days in something like a coma, only recovering consciousness on the evening of the 19th. In fact he appears to have been in a state of shock and total exhaustion, for which prolonged rest was the only treatment available. He resumed his work on the 20th and two days later was well enough to read the last of his lectures on demotic to the Academy of Inscriptions. Meanwhile Jacques-Joseph helped him to prepare a report on his startling breakthrough for the next session of the Academy, which had to be submitted in advance in written form so that lithograph copies could be distributed to those attending the meeting. Word spread amongst scholars that something unprecedented was to happen, and on a dark, wet Friday morning, 27 September 1822, eminent academics of all disciplines began to pack into the Academy of Inscriptions. Tension mounted while they sat through papers given by several prestigious researchers, including de Sacy and Jomard, and with supreme irony Champollion, having never met his closest rival, sat next to Thomas Young who happened to be visiting Paris at that time. Having attended the Academy of Sciences earlier that week, Young had heard what was to take place and was about to witness first-hand the revelation of some of the main principles of the hieroglyphic script and the first stage in the shattering of his own dreams.

The principles that Champollion had suddenly understood, and on which he was to build in the following months and years, were that phonetic hieroglyphic signs were not just restricted to foreign names during the Greek and Roman periods but were used widely in earlier Egyptian writing as well. In reality, as he was later to establish, the hieroglyphic system of writing was based on three main types of sign: pictograms (which are sometimes regarded as a particular form of ideogram), ideograms and phonetic symbols, as well as signs used in special ways such as determinatives. The intricacy of the system stems from the fact that a single sign can often function in more than one way. For example, the sign ⸱ can be used as a simple pictogram, signifying the actual word being represented, so that this picture of a duck means 'a duck,' but it can also function as an ideogram. As such, ⸱ conveys the meaning 'son of,' and is commonly found in the title ⸱ 'sa-Ra' meaning 'son of the god Ra,' which frequently precedes the names of pharaohs. The third way in which the sign ⸱ can be used is phonetically, to represent the sound 'sa,' as in the word ⸱ which is *saw*, meaning 'wooden beam.' Years later, Champollion gave a succinct definition of hieroglyphic writing: 'It is a complex system, a script at the same time figurative, symbolic and phonetic, in the same text, phrase, I would almost say in the same word.'

Although he discovered the underlying phonetic principle of the hieroglyphic writing system, Champollion consistently misinterpreted one of the signs he first saw in the Abu Simbel cartouche: ⸱ actually means 'ms' (usually spelled 'mes'), rather than simply 'm.' It was the German Egyptologist Richard Lepsius (coincidentally born twenty years to the day after the birth of Champollion) who detected this error. In the case of ⸱, the name appears to be literally 'Thoth-mes-s,' but the final 's' is a sign called a phonetic complement, whose function is to make clear that the sign before it should end in 's.' Phonetic complements are frequently added to biliteral hieroglyphs (which have two consonants, such as ⸱ representing 'ms' or 'mes') and to triliteral hieroglyphs (which have three consonants, such as ⌐ representing 'ntr' or 'neter').

The hieroglyphs that are used as phonetic complements are the uniliteral hieroglyphs, such as ⎮ 's' and ▢ 'p,' which represent single consonants – the same hieroglyphs that Champollion had first of all recognized in non-Egyptian names such as Cleopatra. One of the modern methods of classifying phonetic hieroglyphs is according to the number of their consonants: uniliteral, biliteral and triliteral. The Egyptians had no such concept as an alphabet, but the uniliteral signs were used like alphabetic letters and were the most common signs. There were twenty-four in all, although two represent weak consonants or semi-vowels.

| Sign | Object Depicted | Approximate Modern Sound |
|------|-----------------|--------------------------|
|  | vulture | a cross between an *a* and an *h* |
|  | flowering reed | *i* or a weak *y* |
|  | flowering reeds | *y* or *ii* |
|  | forearm | guttural sound similar to *a* |
|  | quail chick | *w* or *u* |
|  | foot | *b* |
|  | stool | *p* |
|  | horned viper | *f* |
|  | owl | *m* |
|  | water | *n* |
|  | mouth | *r* |
|  | reed shelter | soft *h* |
|  | wick of twisted flax | harsh *h* |

| Sign | Object Depicted | Approximate Modern Sound |
|---|---|---|
| ⊜ | placenta (?) | harsh *ch* (as in Scottish *loch*) |
| ⌐ | animal's belly with teats | soft *ch* (as in German *Ich*) |
| ∫ or ⌐ | folded cloth or door bolt | *s* |
| ⌐ | pool | *sh* |
| ◿ | hill slope | *q* or *k* |
| ⌣ | basket with handle | *k* |
| ⊠ | jar stand | hard *g* |
| ⌒ | loaf | *t* |
| ⇒ | tethering rope | *tj* or *tsh* |
| ⇒ | hand | *d* |
| ⇃ | snake | *dj* |

Many of these uniliteral alphabetic hieroglyphs were used to spell out the sounds of foreign names, such as ▯ for 'P' and ⌒ for 't' in Ptolemy, which can give an idea of their original pronunciation. Sounds unfamiliar to the Egyptians were spelled out in an approximate way. For example, the biliteral ᚠ with a sound like 'ua' (or perhaps 'wha' as in 'what') was used for the 'o' sound in Ptolemy, and ᛗ probably pronounced as 'ba' was used for 'B' in Berenice. Champollion's favourite sign ⚞, the recumbent lion used for 'l' in Ptolemy, was actually a biliteral with a sound like 'ru' that was often used to spell out 'l,' a consonant which does not occur in ancient Egyptian.

At the Academy of Inscriptions and Literature meeting of 27 September, Champollion was finally called to give the principal paper, his report on phonetic hieroglyphs, which immediately afterwards he expanded into a publication formally addressed to Bon-Joseph Dacier, the Perpetual Secretary of the Academy. This report, which became a landmark in Egyptology, is now known simply as the 'Letter to Monsieur Dacier.' What seems strange is that although it was his discovery of the names of Ramesses and Tuthmosis, in the drawings from Abu Simbel, which led him to the certainty that phonetic hieroglyphs were first used before the Greeks and Romans, he did not mention this in his Academy talk or in the subsequent publication. Champollion only suggested rather than demonstrated that phonetic hieroglyphs might have been used earlier than the Greek period in Egypt: 'I therefore think, Monsieur, that *phonetic* writing existed in Egypt in quite a far-off time; that it was firstly a necessary part of ideographic writing; and that one then also used it, as was done after Cambyses, to transcribe in the ideographic texts (roughly it is true) the proper names of foreign people, countries, towns, kings and individuals.' If he was not yet prepared to explain his system in full, Champollion did at least reveal many of his results which had been confirmed by his discoveries on the morning of 14 September, a number of which had been worked out while he was still in Grenoble. It was the new hieroglyphic texts from Abu Simbel which provided the proof that at last he was on the right track. After so many years of work, often on the edge of desperation, the sudden shock of success had triggered a complete collapse. Now convinced of the accuracy of his results he set out the phonetic hieroglyphs that were used to spell the names of the Greek and Roman rulers of Egypt. He also gave many examples of those names in their hieroglyphic form along with their translations to demonstrate that his system worked, names such as Alexander the Great, countless Ptolemies and Cleopatras, Tiberius, Trajan and Hadrian. Yet even now Champollion maintained that for hieratic and demotic writing he hoped 'to have succeeded in demonstrating that these two types of writing were, each of them, not alphabetical...but *ideographic*, like the very hieroglyphics [as opposed to his phonetic hieroglyphs], that is to say painting *ideas* and not *sounds* of a language,' showing that he was still uncertain about the relative roles of ideograms and phonetics.

Later scholars have speculated on the reasons why Champollion did not go further in revealing his findings, but only a week had elapsed between his sudden realization of how the hieroglyphs worked and finishing the text of his report on 22 September, so that it could be printed in time for the meeting on the 27th. Still suffering after-effects from his collapse, as well as preparing and giving a further talk on demotic, there had not been time to explore the implications of his discovery, and he was still extremely wary about making his findings public prematurely. As it was, his table of phonetic hieroglyphs and the decipherment of so many names from Greek and Roman Egypt were enough to cause amazement among the academics, and those who gathered to congratulate him included de Sacy, who was now his friend and supporter, and Young, his closest rival.

Champollion's success was based on twenty years of obsessive hard work, all too often in difficult circumstances, and he would soon be able to read the literature from 3,000 years of human history that had been unintelligible for centuries. It was thought of such importance that the King was immediately informed of the discovery, and the newspapers did not take long to report what had happened – with the Dendera zodiac still fresh in people's minds, the story was a sensation throughout Paris as well. Although few had noticed it in the excitement of the other revelations, Champollion's report also settled the argument of the date of the Dendera zodiac by making public what he had previously withheld from his letter published in the *Revue encyclopédique*. He now demonstrated that the name in the cartouche originally alongside the Dendera zodiac was 'Autocrator,' the Greek word for 'Emperor' that was used in the Roman period – Young had earlier translated this cartouche wrongly as 'Arsinoë,' having arrived at this interpretation by his misguidedly mechanical methods. In the letter to Dacier, Champollion wrote, 'the cartouche whose reading I have just given establishes, in an incontestable way, that the relief carving and the circular zodiac were carved by Egyptian hands under the rule of the

Romans.' If it had not been eclipsed by his other achievements, this in itself would have caused a sensation because it put an end to the controversy raging among scholars about the date of the zodiac, removing the possibility that the zodiac might be thousands of years old and thus a direct challenge to the Biblical dating of the creation of the world.

After the meeting Young was formally introduced to Champollion by their mutual friend, the astronomer Arago, and the next morning Young called on Champollion in the rue Mazarine, where he found an excited crowd of people. It is clear from his letters that up to this point he had not fully realized the extent to which his 'young assistant,' as he thought of Champollion, had succeeded in deciphering the hieroglyphs. While inwardly believing that it was all due to his own work in the field, Young was initially generous in his praise, even in private letters to his friends. He wrote to William Hamilton, who was now a minister at the Court of Naples:

> If he [Champollion] did borrow an English key, the lock was so dreadfully rusty, that no common arm would have had strength enough to turn it...Beginning with the few hieroglyphics to which I had assigned a 'phonetic' signification, he found reason to conclude that, in the days of the Greeks and Romans at least, a considerable number of different characters were employed for expressing hieroglyphically the letters composing a foreign proper name...You will easily believe that were I ever so much the victim of the bad passions, I should feel nothing but exultation at Mr. Champollion's success: my life seems indeed to be lengthened by the accession of a junior coadjutor in my researches, and of a person too, who is so much more versed in the different dialects of the Egyptian language than myself. I sincerely wish that his merits may be as highly appreciated by his countrymen and by their government as they ought.

In the meetings between them in the days that followed, relations were cordial between the two rivals, and Champollion
generously showed Young many of his documents and took time to
copy out for him parts of the Casati papyrus that were written in
demotic (Young's enchorial) – the same papyrus that had so fortui-
tously given Champollion the name of Cleopatra, and of which
Young noted: 'it was the first time that any intelligible characters,
of the enchorial form, had been discovered among the many manuscripts and inscriptions that had been examined.' In the company
of Arago, Champollion also visited Young in his lodgings and was
introduced to his wife. With promises to exchange information, the
friendliest of relations seemed to have been established between
them, but it was not to last. In a letter Young wrote shortly afterwards to his friend Hudson Gurney (who was by then a Member of
Parliament, as well as vice-president of the Society of Antiquaries of
London), he said:

> Champollion, the author of the book you brought over, has
> been working still harder upon the Egyptian characters. He
> devotes his whole time to the pursuit, and he has been won-
> derfully successful in some of the documents that he has
> obtained... How far he will acknowledge every thing which he
> has either borrowed or might have borrowed from me, I am
> not quite confident; but the world will be sure to remark, *que
> c'est le premier pas qui coûte* [It's the first step that counts] –
> though the proverb is less true in this case than in most oth-
> ers, for here every step is laborious. I have many things that I
> should like to show Champollion in England, but I fear his
> means of locomotion are extremely limited, and I have no
> chance of being able to augment them.

Young was beginning to feel that, irrespective of all his own misi-
dentifications, the few hieroglyphs that he had correctly identified
entitled him to claim the lion's share of Champollion's success.

Blissfully ignorant that the seeds of Young's future hostility were already beginning to germinate, Champollion was overjoyed at his success and at the recognition it had brought him, writing excitedly to Thévenet in Grenoble: 'The lecture that the Institute [its Academy of Inscriptions] wanted to hear has been a complete success. My discoveries on the hieroglyphs have been unanimously judged incontestable; and I have received compliments higher than the towers of Notre Dame.' To his brother-in-law André Blanc, who was curator at the Royal Library, Champollion wrote that his discovery produced astonishment and forced applause and praise, even from all those people who had distanced themselves from him because of their political allegiance.

With the acclaim of the members of the Academy of Inscriptions still echoing in his ears, and confident that the recognition of his achievement would secure his future, Champollion spent the next few weeks preparing his report for proper publication. He recorded his continuing elation in a letter to Thévenet: 'Everyone repeats to me that one of the first vacant places at the Academy will be for me. I am actually beginning to believe with good faith that it will happen. The obstacles and obstructions that I have had to fight have finally just been flattened by the great blow which I have struck!' His report to the Academy was published at the end of October 1822 under the title '*Lettre à M. Dacier relative à l'alphabet des hiéroglyphes phonétiques*' (Letter to M. Dacier relating to the alphabet of phonetic hieroglyphs) in the form of a booklet of forty-four pages and four illustrated plates, printed by Firmin Didot, the foremost publisher in France and printer to the King.

The euphoria of Champollion was matched by the increasing disillusionment and despondency of Thomas Young. After their meetings in Paris, Champollion sent him two copies of the recently published '*Lettre à M. Dacier,*' accompanied by a cordial letter, and exchanges of information and friendly letters followed over the next few months, but while maintaining civil relations with his rival, Young's jealousy had not ceased to fester. He was especially aggrieved because he felt that he had not received proper credit in Champollion's publication, where he is mentioned just twice – the first time in relation to the demotic inscription of the Rosetta Stone, whose study was said by Champollion to be 'due first of all to...M. Silvestre de Sacy, and then...to the late Akerblad and to M. Doctor Young.' A second mention comes a little further on when Champollion discusses the name of the Queen Berenice, which Young had already published. Noting that the letter 'B' was represented by the sign for a type of dish called a patera (although it is now thought to represent a smoking bowl of burning incense), Champollion added a lengthy footnote about where Young had gone wrong:

It is without doubt through the form of this same sign, which has some analogy with the representation of a basket, that M. Doctor Young was led to recognize the name of *Berenice* in the cartouche which indeed contained it. But this English savant thought that the hieroglyphs which formed the proper names could express whole syllables, that they were therefore a sort of *rebus*, and that the initial sign of the name of Berenice, for example, represented the syllable BIR which means *basket* in the Egyptian [Coptic] language. This point of departure altered to a very great degree the phonetic analysis which he tried on the names of *Ptolemy* and *Berenice*, where he nevertheless recognized the phonetic value of four signs: these were the P, one of the forms of the T, one of the forms of the M, and that of the I; but the whole of his syllabic alphabet, established on these two names only, could not at all be applied to numerous phonetic proper names inscribed on the monuments of Egypt. However, M. Doctor Young has done in England on the written monuments of ancient Egypt, work analogous to that which has occupied me for so many years; and his research on the intermediary [demotic] and the hieroglyphic text of the Rosetta inscription, as on the manuscripts which I have recognised as *hieratic*, present a series of very important results.

Champollion then referred to Young's *Encyclopaedia Britannica* article, but made no further reference to him. Scant praise for Young, but at least his unfortunate error in identifying Autocrator as Arsinoë was not mentioned when Champollion discussed that cartouche in relation to the Dendera zodiac.

Young made no attempt to hide his anger at the minimal way in which he had been acknowledged for what in his view were his own crucially important discoveries, and so less than a month after Champollion's triumph at the Academy of Inscriptions, Young's friends suggested he should write a popular book on hieroglyphs, not anonymously like most of his previous work, but under his own name to enhance his own reputation. At first he refused to consider it, but at that time an event occurred that was to change his mind – the excitement of the discovery of an important new papyrus. George Francis Grey, a traveller and Fellow of University College, Oxford, returned to England from Egypt and loaned Young a box containing numerous papyri, all of which had been purchased from an Arab at Thebes in Upper Egypt nearly four years previously. As Young began to sort through them, he examined the only papyrus that was written in Greek and stumbled on an incredible coincidence – this was a Greek translation of the actual Casati papyrus written in demotic (the late Egyptian language) where Champollion had found the name 'Cleopatra.'

I could not, therefore, but conclude, that a most extraordinary chance had brought into my possession a document which was not very likely, in the first place, ever to have existed, still less to have been preserved uninjured, for my information, through a period of nearly two thousand years: but that this very extraordinary document should have been brought safely to Europe, to England, and to me…would, in other times, have been considered as ample evidence of my having become an Egyptian sorcerer.

Apart from Drovetti's worn bilingual stone languishing at Livorno as part of the collection which the King of Sardinia-Piedmont was now negotiating to purchase, this was the first time since the discovery of the Rosetta Stone that a demotic and a Greek version of the same text had been recognized. The contents of the papyrus were no less extraordinary, being concerned with the sale by one cemetery priest to another of a share of the services performed for certain mummies, and in an exultant mood Young wrote to his friend Hudson Gurney in late November 1822 that 'I have already obtained a little miniature triumph over Champollion.' Of Drovetti's stone, he declared: 'if Drovetti's black stone goes to the bottom of the gulf of Genoa I shall care very little about it. I would now scarcely give ten pounds for it.'

Unaware of Young's simmering anger, Champollion was eager to build on his results and was busy studying every example of hieroglyphs that he could lay his hands on. Initially he concentrated on the names of rulers, not only because they could be easily identified by their surrounding cartouches, but because a correct sequence of rulers would provide an invaluable framework around which a history of Egypt could be constructed. The rulers of ancient Egypt were kings, but they were also considered as part human and part god, and their role was inextricably bound up with Egyptian religion. It was believed that the king was of divine birth, a representative of the gods on earth while he lived, becoming a god when he died. Nowadays the kings of ancient Egypt are known as 'pharaohs' – a term derived from a Greek translation of the hieroglyphs ⌐⌐ (per-aa – great house). This can mean 'great house' in the sense of 'palace,' but it came to be used in the idiomatic sense of 'great ruling house' (as the modern monarchs of Britain belong to the 'House of Windsor'), and by 1500 B.C. the term ⌐⌐ came to refer to the king himself.

The cartouches that enabled Champollion to pick out the names of the pharaohs within blocks of text derived their name from the soldiers of Napoleon's Egypt expedition, who thought that these hieroglyphic signs resembled the profile of their gun cartridges –

*cartouche* is the French word for 'cartridge.' The cartouche hieroglyph ⬭ developed from the hieroglyph ☉ (all that which is
encircled by the sun), changing from a circular to an oval shape to
accommodate the number of hieroglyphs needed to represent pharaohs' names. The cartouche sign actually represented a loop made
from a double thickness of rope with the ends bound together to
form a continuous circuit, and the Egyptian word for cartouche
was ⟨shenu symbol⟩ (*shenu*), derived from the word 'to encircle.' Originally it
may have denoted that the person whose name was contained
within the cartouche was ruler of everything encircled by the sun.
Both the signs ☉ and ⬭ were symbols of eternity, and when used
around a name the cartouche sign effectively became an amulet,
providing protection for whoever was named. Their name was an
integral part of a person, so that if a person's name was no longer
written anywhere, it was impossible for them to survive in the afterlife. Names were at times obliterated in an attempt to destroy
someone, and 'loss of name' was one of the penalties for high treason in ancient Egypt. To write the name of the pharaoh inside a
cartouche was both a religious and a magical act to protect the
pharaoh and ensure he would live for ever – a much stronger senti-
ment than the more modern cry of 'Long live the King!'

Champollion was already aware that cartouches often occurred
in pairs, referring to the same pharaoh, and that some cartouches
contained not only the names of the pharaoh, but some of his honorific titles as well. Young was misled into thinking that pairs of
cartouches enclosed the name of the king and then the name of his
father, and he failed to appreciate that groups of cartouches gave
only the name of the ruler with his titles. In fact, the names and
titles of the pharaohs evolved until some time after 2000 B.C. when
each pharaoh came to have a unique combination of five names,
some of which were themselves made up of several elements, and
just two were contained in cartouches. The pharaoh's Birth Name
(sometimes called 'nomen'), the name given to him at birth, was
contained within a cartouche preceded by the hieroglyphs ⟨symbol⟩ (Sa-

Ra) meaning 'Son of the god Ra' to emphasize the pharaoh's divine origin. His other name within a cartouche was his Throne Name (sometimes called prenomen), which he took when he succeeded to power, a name that was preceded by the hieroglyphs ⳨⳨ (*nesu-bity*), literally 'he of the sedge and the bee.' This has the meaning of 'King of the Dualities,' a title with a range of complex interpretations which reflect the stark contrasts that characterize Egypt, such as farmland and desert, and so it is usual to concentrate on the political duality of Upper and Lower Egypt (since the sedge is the symbol of Upper Egypt and the bee the symbol of Lower Egypt) and translate the hieroglyphs as 'King of Upper and Lower Egypt.' Three other names, which were also given to the pharaoh when he took the throne, are actually honorific titles emphasizing his power and divinity rather than real names. These were the Horus Name, preceded by the hieroglyph for the god Horus, ⳨; the Nebti Name (sometimes called the 'Two Ladies Name'), preceded by the hieroglyph ⳨, the sign for the goddesses Nekheb of Upper Egypt and Wadjet of Lower Egypt; and the Golden Horus Name, preceded by the hieroglyph ⳨ (*Hor nebw* – Golden Horus), the sign for Horus of Gold.

For pharaohs reigning after this system was fully developed, the full set of five names could be quite extensive, such as the full names of Tutankhamun, which were normally inscribed in the following order:

⳨ – Horus Name: Ka-nakht tut-mesut, meaning 'Strong bull, fitting of created forms.'

⳨ – Nebti Name: Nefer-hepu segereh-tawy sehetep-netjeru nebu, meaning 'Dynamic of laws, who calms the two lands, who propitiates all the gods.'

⳨ – Golden Horus Name: Wetjes-khau sehetep-netjeru, meaning 'Who displays the regalia, who propitiates the gods.'

[hieroglyphs] – Throne Name: Nebkheperure, meaning 'The
  lordly manifestation of the god Ra.'

[hieroglyphs] – Birth Name: Tutankhamun heqa-iunu-
  shema, meaning 'Living image of the god Amun, ruler
  of Upper Egyptian Heliopolis.'

The birth name of Tutankhamun is a special case, since this one
was not given to him at his birth; instead he was given the name
[hieroglyphs] Tutankhaten (Living image of the god Aten) –
at that time the 'heretic' Pharaoh Akhenaten was ruling, who
only worshipped the god Aten rather than the whole panoply of
Egyptian deities. Tutankhaten's name was later changed to
Tutankhamun.

The combination of the five names was unique to each pharaoh,
but in practice their two names in cartouches are usually sufficient
to distinguish between pharaohs, even when they shared the same
Birth Name. For example, the Pharaoh Tuthmosis I who ruled
Egypt around 1500 B.C. had the Birth Name [hieroglyphs] (Thothmes
– Born of the god Thoth), which is exactly the same as the Birth
Name of the Pharaoh Tuthmosis II, who succeeded him. However,
the Throne Name of Tuthmosis I was [hieroglyphs] (Akheperkare –
Great is the soul of the god Ra), whereas the Throne Name of Tuth-
mosis II was [hieroglyphs] (Akheperenre – Great is the form of the
god Ra). Although the difference between [hieroglyphs]
and [hieroglyphs] is slight, it is enough to distinguish
Tuthmosis I from Tuthmosis II.

To develop his system and go significantly beyond what the cartouches and their related texts could tell him, Champollion needed
as wide a range of hieroglyphic texts as possible, and he was only
too well aware that much of the best material, and indeed the
majority of material, was finding its way to England rather than
France. The British Consul in Egypt, Henry Salt, was acting as an
agent for exporting Egyptian antiquities to England, selling them to
the British Museum and to private collectors. The French Consul in

Egypt, Bernardino Drovetti, also acted as an export agent but he could not find a market in France for the antiquities he had acquired, and his important collection, which was still in store in Livorno where it had been seen by Young in 1821, was in the process of being sold to the King of Sardinia-Piedmont. In England, many of the antiquities with hieroglyphic texts on them, including many papyri, were now passing through the hands of Young before disappearing into private collections – ironically, while Young had a surfeit of material that he was unable to read, Champollion was ransacking Paris for any hieroglyphic texts that he had not seen before. The final insult was that one of the most important collections was being amassed by Young's friend William Bankes, who spitefully refused to let Champollion see any copies of his material, perhaps in support of Young or simply because he disliked the French or because of the way the text on his own obelisk with the cartouche of Cleopatra had been successfully used by Champollion.

Such was the situation that led to Champollion being in a Paris sale room early one morning in January 1823, hurriedly copying the hieroglyphic texts on some of the sale items before the public arrived. His rapid, confident copying of the hieroglyphs was observed by one of the gentlemen arriving for the sale, who engaged Champollion in conversation about Egyptian collections, and in particular the collection of Drovetti, which appeared to be on the point of going to Turin in Italy rather than France. Sensing a kindred spirit, Champollion abandoned all caution and passionately put forward how important Drovetti's collection was, both for its historical content and for advancing knowledge of the ancient Egyptian language. He complained bitterly that the French government had paid 150,000 francs for the Dendera zodiac, but would not pay for Drovetti's collection.

Despite the sensation it had caused in Paris, the Dendera zodiac was trivial compared with Drovetti's collection, but only those with expert knowledge of the subject were aware of this. Taken aback at Champollion's forthright and knowledgeable summary, which reflected his own views, the gentleman revealed himself as the Duke of Blacas. Apart from having a deep interest in archaeology, particularly that of the East, Pierre Louis Jean Casimir Blacas d'Aulps was an aristocrat and loyal Royalist. Although politically very different from Champollion, he nevertheless held liberal views and had many enemies among the Ultra-Royalists who had caused Champollion so much pain and anguish. Following the Revolution Blacas had left France, served as a soldier and found himself exiled in England with the future King Louis XVIII. With the restoration of Louis XVIII, he was given the post of First Gentleman of the Chamber of the King, and Louis relied on his advice: Blacas was one of the most influential men in France.

This chance meeting was a turning point for Champollion, who made such an impression that Blacas promised his support. He was not slow to mention Champollion to the King, and despite Champollion's past political record and the fact that his great discovery had not been dedicated to Louis XVIII, but to Dacier, the King was persuaded to acknowledge his achievement with the gift of a gold box engraved with the inscription 'The King Louis XVIII to M. Champollion on the occasion of his discovery of the Alphabet of the hieroglyphs.' Blacas delivered the box to Champollion along with a strong hint that it would be as well for further findings to be put under the protection of the monarch with a suitable dedication.

The euphoria that had transformed Champollion's life following his sudden realization of the principles of the hieroglyphic script began to fade as he came to understand that praise for his scientific achievement was not going to be matched by an immediate improvement in his circumstances. Without a paid position, he was still suffering from poverty and fragile health – it seemed that all the promises of fame and fortune had been only window-dressing. The support of Blacas came just in time for Champollion, whose enemies were beginning to recover from the shock of his revelations and were finding ways to attack or belittle him. Jomard, who continued to edit the *Description de l'Égypte*, was his main opponent in

France. Seldom acting openly, he insinuated through his acquaint-ances among the academics in Paris that Champollion had not really achieved anything and that the hieroglyphs still remained to be deciphered. He used everything he could to undermine Champollion's credibility, pointing out to anyone who would listen that Champollion had never been to Egypt. Several other rivals took a similar view to Jomard – unable to accept that Champollion had succeeded where they had failed, they convinced themselves that the problem had not been solved and regarded him as a charlatan.

Almost worse than his enemies were some of his new allies. Having demonstrated that the date of the Dendera zodiac was much more recent than had previously been believed, and therefore that it did not provide evidence that the Biblical chronology was flawed, Champollion had not only increased the enmity of Jomard, who maintained that the zodiac was many millennia old, but he had also won the admiration of many priests of the Church as a defender of Catholic dogma. Because Champollion had approached the problem of the date of the Dendera zodiac as a scientific investigation, he was irritated by the religious bigotry that surrounded it, and he complained to his old friend Augustin Thévenet that he was sick of being regarded as a 'Father of the Church' and a 'bastion of the faith' and was weary of the 'odour of sanctity' that surrounded him.

While aware of many of his enemies and unwelcome new allies in France, Champollion still thought that Young in England was reconciled with his views, but in this he was badly mistaken. Early in 1823 the journal *Quarterly Review* carried an anonymous review of his '*Lettre à M. Dacier,*' which attributed to de Sacy the discovery of the relationship between hieratic and hieroglyphic writing, to Åkerblad the beginnings of the hieroglyphic alphabet, and to Champollion merely the development of Young's hieroglyphic alphabet. The review had been written by Young, and Champollion recognized it as such. At the same time the journal announced a forthcoming publication by Young with the provocative title of *An Account of some recent discoveries in hieroglyphical literature and Egyp-*

*tian Antiquities, including the author's original alphabet, as extended by*
*Mr. Champollion.*

Outraged and hurt, Champollion wrote to Young from Paris on
23 March 1823, refuting the claims of this anonymous writer
while maintaining the pretence that Young was not responsible:

I have just read... the analysis of my 'Lettre à M. Dacier,' on my
alphabet of Phonetic hieroglyphs. It has produced in me the
same effect as on everyone else who has read it, and who has
cried out greatly against the ignorance or the bad faith of the
author of this article. The facts relating to my alphabet are too
well known, too public; and the times of the attempts made on
this subject by various savants are too well fixed, that one
should not fairly condemn the thoughtless assertions of the
author of this article, who tries to give to others what evi-
dently belongs to me. Nobody has understood here, and M. de
Sacy less than any other, how the discovery of the relationships between demotic writing and what I have called hieratic
with hieroglyphs has been attributed to him, as he has never
been occupied with this in his published work. Nobody under-
stands the assertions about M. Åkerblad on the hieroglyphic
text of Rosetta, nor the names that are claimed to have been
read with the help of his alphabet, either in other hieroglyphic
inscriptions, or in other manuscripts on papyrus. Even more
absurd assertions which are in this same article are equally
not understood... As for the claimed uselessness of my discovery in regard to its application to the general system of
hieroglyphs, the Academy already knows where it stands,
and very shortly the literate public will be as convinced that
my alphabet is the true key of this whole system. I find in the
same journal the announcement of a volume which you are
about to publish and whose title claims to show the true
author of an alphabet which I have done nothing but to
extend. I shall never consent to recognize any original alpha-

bet other than my own, where it is a matter of the hieroglyphic *alphabet* properly called; and the unanimous opinion of the savants in this regard will be more and more confirmed by the public examination of all other claims. I am therefore going to reply to the anonymous author of the fore-going article...I do not believe that you accept the claims of the anonymous author; and my estimation of your character is too profound for me to have hesitated for one instant to share my thoughts with you on this subject.

Although Young pretended he had not written the anonymous review, it was clear to all those not biased in his favour (and even to some who were) that Champollion had a system of decipherment, whereas Young had only translated a few signs, of which some were correct and some were not. Shortly after the *Quarterly Review* announcement, Young's book was published, provocatively laying claim to the discovery of the hieroglyphic alphabet. Even though he wrote in his own name for once, he did not hesitate to use his book as an outpouring of his resentment. He believed that this French-man Champollion had succeeded by using the substantial foundations that he himself had built, without so much as a proper acknowledgement, let alone the acclaim that he felt was his due:

> It would have indeed been a little hard, that the only single step, which leads at once to an extensive result, should have been made by a Foreigner, upon the very ground where I had undergone the drudgery of quietly raising, while he advanced rapidly and firmly, without denying his obligations to his predecessor, but very naturally, under all circumstances, without exaggerating them, or indeed very fully enumerating them.

However, Young's resentment of the 'foreigner's' success went much deeper, since he believed that Champollion was also building a reputation at the expense of his own. A prime intention in Young's *Account of some recent discoveries* was to publish the newly discovered papyrus that matched the Casati demotic papyrus and at the same time to vindicate his claim to have achieved all the key work on the decipherment of hieroglyphs. He therefore included accounts of his work on the Rosetta Stone as well as on other demotic papyri and the role of Champollion as perceived by him, expressing many points of contention and criticism, but pretending that his own method had been proved: 'his Letter to Mr. Dacier, since printed; in which I did certainly expect to find the chronology of my own researches a little more distinctly stated...But, however Mr. Champollion may have arrived at his conclusions, I admit them, with the greatest pleasure and gratitude, not by any means as superseding my system, but as fully confirming and extending it.' Young had convinced himself that Champollion owed much of his work to his own researches and was indignant that Champollion had only acknowledged the discovery of four alphabetic letters by Young, believing he was responsible for discovering nine:

> but instead of *four* letters which Mr. Champollion is pleased to allow me, I have marked, in a subsequent chapter of this Essay, *nine*, which I have actually specified in different parts of my paper in the Supplement: and to these he has certainly added *three* new ones; or *four*, if he chooses to reckon the E as a fourth. I allow that I suspected the B, the L and the S, to be sometimes used syllabically.

Young was obviously unaware that such quibbling over the number of hieroglyphs he had correctly identified, even if it was nine instead of four, only undermined his credibility.

Apart from making him exasperated and angry, Young's publications made Champollion realize that his discoveries were still being vigorously contested and he decided that he should no longer be cautious but must publish more of his findings as soon as possible, and give more lectures to the Academy – decisions that show his increasing self-confidence. He was already working towards a more comprehensive explanation of his system of decipherment than had been set out in his 'Lettre à M. Dacier,' and he had also begun work on a publication about the gods and goddesses of Egypt which was to be made available in a series of booklets under the title *Panthéon égyptien*, subscriptions to which he hoped would earn him some money. The first booklet in the series was published in July 1823, and a further eight booklets appeared by the end of 1824, after which they were published at irregular intervals. They contained beautiful coloured plates, done by his friend Jean-Joseph Dubois, based on the limited number of drawings of deities available to them and accompanied by relevant hieroglyphs. As they were published before the expanded explanation of his system of decipherment, the initial booklets provided a target for criticism from his enemies, who were not always persuaded of the veracity of Champollion's explanations of the gods and goddesses.

In late August 1823 Champollion sent Young the first booklet of his *Panthéon égyptien*, saying in the covering letter:

My aim in publishing this collection is to make clear the various mythical people represented on the monuments of Egypt, to distinguish them one from another; without claiming to enter into the very foundation of their emblematic or symbolical significance. It is simply a *recognition* pushed into the up-to-now inextricable labyrinth of the Egyptian Olympiad. The rest will depend on the true progress that we will make in the hieroglyphic method.

Young's reaction was to damn it with faint praise, writing to his friend Sir William Gell who had lived in Italy since 1820: 'Champollion has sent me the first number of his Pantheon, which must on the whole be an important collection; but he appears to me to be much too hasty; and he does not add to his deities enough of the hieroglyphics found with them to enable one to judge of the name intended to be applied to them.'

Despite his bravado and his continued defence of his own hieroglyphic studies, Young was becoming tired of the subject, and in September he wrote to Gell about his intention to give up publishing on Egyptian subjects for reasons of expense, lack of new material and the fact that 'Champollion is doing so much that he will not suffer anything of material consequence to be lost. For these three reasons I have now considered my Egyptian studies as *concluded.*' This was one of several similar declarations that he was to make over the next few years, but while he gave up serious study of hieroglyphs, he never fully relinquished his bitterness and resentment of Champollion.

Through the rest of 1823 Champollion laboured on the booklets of the *Panthéon égyptien* and on the enlarged and up-to-date explanation of his decipherment system, a book which was to be called *Précis du système hiéroglyphique des anciens Égyptiens* (A Summary of the Hieroglyphic System of the Ancient Egyptians). By December this *Précis* was nearing completion, and through his contact with the King's favourite, the Duke of Blacas, he hoped to be able to present a copy to the King in person. As a remedy for the shortage of suitable hieroglyphic texts, Champollion had also decided that he needed to go to Italy to study both Drovetti's collection, which was now deposited at Turin, and several other collections, such as that at the Vatican in Rome – collections which made Italy the richest source of hieroglyphic texts outside Egypt. Apart from the prudence of obtaining royal patronage for such a trip, Champollion's continued poverty made it essential that he obtain funding as well, but although he was very willing to intercede for Champollion, Blacas himself had political enemies, and when he now fell ill it was rumoured that he had fallen from the King's favour.

With Blacas out of the way, his enemies at Court changed the subscription for a copy of the *Précis* from a luxury edition to a less expensive one, which appeared to be proof to Champollion that

Blacas was indeed in disgrace. It was now a year since the acclaimed announcement of Champollion's discoveries, and with no appreciable difference in his circumstances and with persistent ill health, his optimism was beginning to turn to despair. His mood was not helped by events in Grenoble. His wife Rosine, who was now pregnant, had been there for some time nursing her father who eventually died in January 1824. She had never been given her dowry, and in a dispute between her brothers about inheritances, she appears to have been denied a share of her father's wealth despite support from Jacques-Joseph, who was then also in Grenoble – it was obvious that Champollion's worries about money would not be solved by finances from his wife's family. Depressed and feeling more and more unwell, Champollion now wrote to his brother about the damage done by the strain of preparing his *Précis*: 'my poor head hurts, my tinnitus, the humming and buzzing noises, has worsened and leaves me neither night nor day. I have frequent spasms and am incapable of occupying myself seriously for more than a quarter of an hour. I attribute this worsening to my plates [of the *Précis*] which have forced me to remain bent over for more than a month.' Taking care not to worry his wife, he added: 'It is pointless to speak of this to Rosine; tell her that...things aren't going badly.'

Jacques-Joseph returned to Paris in late January, just in time to accompany his brother on a last visit to Louis Langlès, who had been Champollion's professor of Persian. The dying Langlès had earlier been reconciled with Champollion, who felt that his recent support outweighed the bitterness that had earlier come to exist between them. The parting was a sad one, and Langlès died on 28 January. Fortunately the illness of Blacas was neither life-threatening nor long-lived. His return to Court dispelled all the rumours of his fall from the King's favour, but it was not a simple matter to reverse the actions of his enemies, and so the official publication of the *Précis* was delayed while Blacas sought a suitable opportunity for Champollion to present a copy to the King.

Although the *Précis* had been ready since the middle of January, the presentation was repeatedly postponed, but Champollion was once again full of hope because Blacas seemed to be confident that he could eventually secure the King's backing for the project to study the Egyptian collections in Italy. If funds were not forthcoming from the King, Blacas had even promised to underwrite the project, and so with increasing excitement Champollion made plans and preparations for his forthcoming tour of Italy. Jacques-Joseph was heavily involved with the plans, and both he and Champollion wrote to various friends and acquaintances for information about what best to study and how to proceed in Italy, because Blacas needed a detailed itinerary to put before the King if there was to be any hope of royal finance for the project.

By this time constant practice at deciphering inscriptions had made Champollion very proficient in reading them, and he had the advantage of being fluent in the Coptic language which was the nearest equivalent to the ancient Egyptian language. When deciphering a text he transliterated the hieroglyphs into the Coptic language using the Coptic alphabet, and it was then not too difficult to translate the Coptic text into French. Champollion found that this two-stage process was easier and more accurate than trying both to decipher the hieroglyphic symbols and to translate Egyptian to French simultaneously. However, the system was not perfect because ancient Egyptian is far removed from Coptic, just as English or French is far removed from Latin. It might appear possible to deduce the meaning of the Latin word *plumbum* from the French word *plomb* or the Latin word *lacus* from the English word *lake*, but doubt always exists because all languages change over time. Nowadays Egyptologists still use a two-stage process, but they transliterate hieroglyphs into an alphabet with extra phonetic symbols to cover the sounds not represented in that alphabet, before translating the ancient Egyptian text into another language, such as English or French. For example, ⸗ transliterates as *ḥmt-nṯr*, which means 'priestess.'

All the years Champollion had spent learning Coptic now proved invaluable, and in February he noted his progress in the study of hieroglyphs in a letter to Count Lodovico Costa at Turin: 'All my results are based on the monuments...no longer is a single one silent for me, provided that it carries one of the religious symbols or some Egyptian inscription.' Costa was the ambassador from Sardinia-Piedmont with whom he had worked in Grenoble, and who had offered him the Chair of History and Ancient Languages at Turin, and Champollion was impatient to go to Turin to see the Drovetti collection there. In the same letter, aware that even if he secured financial backing from King Louis XVIII it might not be enough, Champollion asked Costa if his government might pay the expenses of his stay in Turin if he provided a scientific catalogue of the Drovetti collection in return. It was so important for him to see the material in Italy that he would do everything possible to ensure success.

After some weeks of patient observation, Blacas found an opportunity to broach the subject with the King, and on 29 March he finally arranged a meeting between Champollion and Louis XVIII. A long conversation took place and Champollion presented the King with a copy of his *Précis*, but the proposed trip to Italy could not after all be mentioned because rumours about it were already circulating at the Court – Champollion's enemies among the Royalists were again trying to blacken his name and had succeeded to the extent that the King had ordered another investigation into his past. Fortunately Blacas was able to counter the lies and insinuations, and in April he was able to submit to the King the report and supporting notes on the projected trip to Italy that had been prepared by Champollion and Jacques-Joseph – with the result that the King immediately ordered the necessary financial backing to be paid. As a finishing touch to the extreme happiness that this news brought to Champollion, Blacas invited him on a visit to Naples on the assumption that by the time he reached Italy, Blacas would be appointed French ambassador there.

Once the presentation to the King had taken place, Champollion was free to publish his *Précis* and it went on sale in the middle of April 1824. This book went a long way to explaining what he had only asserted in his '*Lettre à M. Dacier*' and it caused a similar sensation. The *Précis* was an incredible volume, crammed full of his recent and wide-ranging discoveries on hieroglyphs. The preface stated that his phonetic alphabet, 'whose first result was to irrevocably fix the chronology of the monuments of Egypt...acquired an even higher degree of importance still, because it became in some way for me what one has commonly called the true *key of the hieroglyphic system*.' Champollion devoted several pages to the progress of decipherment, showing where Young had gone wrong and what he himself had achieved. There followed explanations of the hieroglyphic signs, what they meant and why, discussions on proper names of kings and private people, titles of rulers, the meaning of multiple cartouches, the various types of writing such as hieratic and demotic, non-phonetic hieroglyphs (today called pictograms and ideograms) and an outline of some grammar.

Also in the *Précis* was a discussion of the number of signs: 'the famous Georges Zoëga...succeeded in gathering a range of 958 hieroglyphic signs which he regarded as quite distinct. I am led to believe that this Danish savant often noted as different signs characters which at heart were only inconsequential variations of one another...I have only been able to obtain a numerical result lower than that of Zoëga.' Champollion arrived at a total of 864 signs, but today it is thought that the earliest hieroglyphs numbered about 1,000, decreasing to 750 around 2000 B.C. and increasing to several thousand in Ptolemaic Greek and Roman times.

Scholars' reactions to the *Précis* were divided, with friends and enemies of Champollion taking different sides, but outside this small circle his achievement had become a nationalist issue, not least because the main opposition to Champollion came from England. After all the hardships of the Napoleonic era and subsequent blows to their national pride, the French were eager to seize on any

French achievement, and with the publication of the *Précis* Champollion once again became a celebrity in Paris. His health was still not good, and so Jacques-Joseph did what he could to shelter him, receiving innumerable visits from enquirers and admirers and passing on messages to his brother, who was concentrating on the preparations for his journey through Italy.

Copies of the *Précis* were sent to a long list of people who had helped Champollion, including Louis-Philippe, the Duke of Orleans, whom Champollion had met when both were founder members of the Asiatic Society. Louis-Philippe had come to be one of his supporters and was well aware of the wider implications of his achievement. In his role as Honorary President at the first public meeting of the Asiatic Society a year previously, he had opened the proceedings with a speech in which he acknowledged Champollion's success:

> The brilliant discovery of the hieroglyphic alphabet is honourable not only for the savant who has made it, but for the nation! It must make one proud that a Frenchman has begun to penetrate the mysteries that the Ancients only revealed to some rather experienced followers and to decipher these emblems, whose significance all modern people were desperate to discover.

By May everything was ready for the journey to Italy, but an unusually severe winter in the Alps was still blocking the mountain passes. In Paris the political situation at Court was not stable, and, with the imminent departure of Blacas for Naples, Champollion feared that his enemies might get the upper hand and block the trip. As there was now time and money for a visit to England, he seized the opportunity to travel to London with Jacques-Joseph, primarily to see the collections in the British Museum, but most of all to examine the Rosetta Stone inscription itself, good copies of which had been so difficult for him to obtain for so many years. The trip was necessarily short because of the need to return to France and start off for Italy before Blacas left for Naples, and the only surviving record of Champollion's reaction to England is in a letter written two years later by San Quintino, the Keeper of the Egyptian Museum at Turin. He had sided with the enemies of Champollion, and in an attempt to denigrate him, San Quintino wrote to Young, '...M. Champollion, who, in speaking to me one day of his journey to England, and to the British Museum, told me that the English are barbaric.'

William Warburton, the eighteenth-century commentator on hieroglyphs.

Jean-Jacques Barthélemy, the first to recognize the purpose of cartouches within hieroglyphic inscriptions.

Napoleon Bonaparte at the time of his Egypt Expedition.

The Rosetta Stone after recent conservation, with the bottom left-hand corner remaining unconserved to show the dark wax covering the surface and the white infill.

The Egyptian-style tomb of Joseph Fourier in the Père Lachaise cemetery in Paris: his bust was recently stolen and replaced by one from a neighbouring tomb.

Marie-Alexandre Lenoir, who published a book on hieroglyphs when Champollion was a student at Paris.

The tomb of Edme-François Jomard, in the Père Lachaise cemetery in Paris, in the form of an Egyptian obelisk.

Thomas Young, the most serious rival to Champollion in the decipherment of hieroglyphs.

The birthplace of Thomas Young at Milverton in Somerset.

Jean-François Champollion was born in the house (now a museum) at the end of the rue de la Boudousquerie in Figeac.

Jean-François Champollion (left) and Jacques-Joseph Champollion-Figeac at the beginning of the nineteenth century.

The original entrance to the municipal library and museum in Grenoble where Jacques-Joseph and his brother worked.

Jean-François Champollion in 1823, holding in his hand the table of phonetic signs from his *Lettre à M. Dacier.*

Entrance of the Cour Carrée of the Louvre in 1830, at the time when Champollion was curator of the Egyptian collections.

A painted relief at the entrance of the tomb of Ramesses III in the Valley of the Kings as published in the *Description de l'Égypte.*

Above; Funerary text of the Litany of Ra inside the tomb of Ramesses IV in the Valley of the Kings at Thebes, near Champollion's sleeping quarters.

Below: Looking down the entrance corridor of the tomb of Ramesses IV in the Valley of the Kings which was used as temporary accommodation by Champollion's expedition.

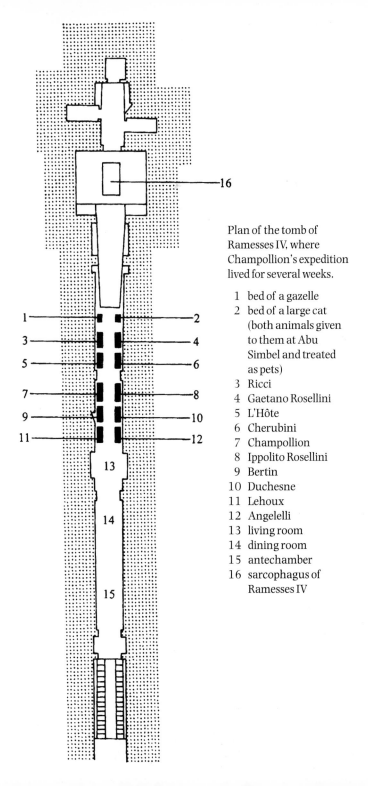

Plan of the tomb of
Ramesses IV, where
Champollion's expedition
lived for several weeks.

1 bed of a gazelle
2 bed of a large cat
  (both animals given
  to them at Abu
  Simbel and treated
  as pets)
3 Ricci
4 Gaetano Rosellini
5 L'Hôte
6 Cherubini
7 Champollion
8 Ippolito Rosellini
9 Bertin
10 Duchesne
11 Lehoux
12 Angelelli
13 living room
14 dining room
15 antechamber
16 sarcophagus of
   Ramesses IV

Cartouche at Karnak of the Birth Name of the Pharaoh Tuthmosis IV (ruled around 1419 to 1386 BC).

Formulaic hieroglyph from Karnak: the ↾ (was), 𓊽 (djed), ☥(ankh) and ⌣ (neb) signs are frequently combined in this way, meaning 'all power, stability and life'.

Columns of hieroglyphs on a pylon wall at Karnak containing texts of the Pharaoh Amenhotep III (ruled around 1386 to 1349 BC).

Different types of Egyptian writing.

1 hieroglyphs
2 hieratic
3 demotic
4 Coptic

A table of 'pure' and linear phonetic hieroglyphs with hieratic and demotic equivalents from Champollion's *Précis du système hiéroglyphique*.

Cartouches of pharaohs from Champollion's *Précis du système hiéroglyphique*.

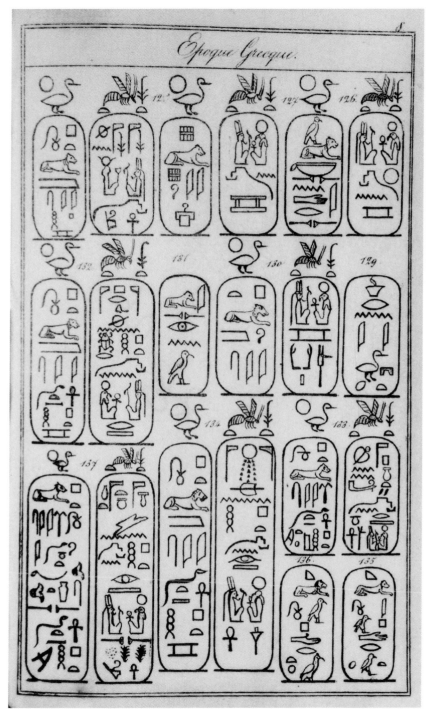

Cartouches of Greek rulers of Egypt from Champollion's
*Précis du système hiéroglyphique.*

The Egyptian-style obelisk in Figeac commemorating Jean-François Champollion.

Jean-François Champollion's tomb in the form of an Egyptian-style obelisk in the Père Lachaise cemetery in Paris.

Map of Europe and Egypt.

| | |
|---|---|
| 1 Figeac | 9 Florence |
| 2 Grenoble | 10 Rome |
| 3 Paris | 11 Naples |
| 4 London | 12 Paestum |
| 5 Edinburgh | 13 Toulon |
| 6 Göttingen | 14 Alexandria |
| 7 Turin | 15 Cairo |
| 8 Livorno | |

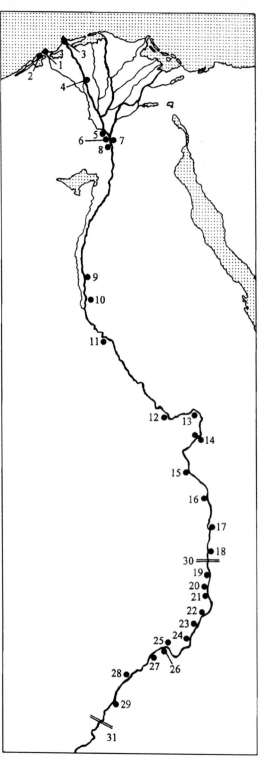

Map of the Nile Valley
of Egypt and Nubia.

1  Aboukir
2  Alexandria
3  Rosetta
4  Sais
5  Embaba
6  Giza
7  Cairo
8  Saqqara
9  Beni Hassan
10  El-Amarna
11  Assiout
12  Abydos
13  Dendera
14  Thebes-Luxor-Karnak
15  Esna
16  Edfu
17  Kom Ombo
18  Aswan/Philae
19  Kertassi
20  Beit el-Wali
21  Kalabsha
22  Girf Husein
23  Dakke
24  Wadi el-Seboua
25  Amada
26  Derr
27  Qasr Ibrim
28  Abu Simbel
29  Wadi Halfa
30  First Cataract
31  Second Cataract

CHAPTER EIGHT

# (Master of Secrets)

Back in Paris at the end of May 1824 Champollion decided it was best to leave secretly for Grenoble and wait there until the Mount Cenis Pass across the Alps reopened, hoping that he would be in Italy before his enemies at Court noticed he had gone. On his arrival in Grenoble he was welcomed as a hero by his old friends, but, exhausted after two days of their constant conversation, questions and celebrations, he retreated to Jacques-Joseph's house at Vif, just south of Grenoble. Here, in the house run by his sister-in-law Zoé, Champollion was reunited with his wife Rosine and saw for the first time his daughter Zoraïde, who had been born on 1 March while he was still trapped in Paris, obliged to be ready each day for word from Blacas about his presentation to the King.

For a few idyllic days in Vif he could relax with his family, free from all pressures except one – Egypt, with its history and its hieroglyphic texts, was never completely banished from his thoughts. The holiday lasted barely a week before news came that the passes were opening. Setting out for Italy on 4 June he reached Turin three days later after a journey through the Alps on what he regarded as excellent roads, but, as he wrote to his brother, ones laid out 'on the edge of terrifying precipices, and there is never enough daylight by which to guide a heavy diligence,' so that descending slopes in such a cumbersome coach always brought to his mind the danger of the driver losing control. Count Costa, now Secretary of State for

Sardinia-Piedmont, could not be in Turin to greet him, but Champollion was warmly welcomed and given temporary accommodation, and by 10 June Costa had arrived and insisted that he lodge with him. By this time too the official authorizations for Champollion to work on the Drovetti collection had arrived, and he was more than ready to realize his dream in the museum at Turin.

In a letter to Jacques-Joseph, he recorded his first impression of the amazing Drovetti collection, which was beyond anything he had imagined: 'I will tell you in one phrase of the country: *Questo e cosa stupenda*! [That is an incredible thing!]' Rooms were filled with colossal pieces of sculpture carved in green, grey, black and rose-coloured granite, all finely polished, but the inscriptions on these pieces were surpassed by the magnificent collection of papyri. As Champollion conducted a rapid appraisal of the collection, he became aware of the tremendous task that lay ahead. In front of him was a huge number of texts that he found he could decipher with increasing ease – texts that spoke to him of unknown kings, strange funerary rituals, diplomatic correspondence and letters written by the inhabitants of ancient Egypt, just like the letters he himself habitually wrote to his family and friends. It was only one collection – just a few grains of sand in the desert – what of the other collections in Italy and elsewhere, and what about Egypt itself? The task that presented itself obviously could not be completed in a single lifetime – it was overwhelming – and a great deal of hard work would be needed just to make an impression on the wealth of information that was on offer. Before he left Turin, Champollion was to write about the importance of examining the original monuments rather than relying on other people's inaccurate copies:

I see the route that I must follow, I know the means which are still available to use in order to advance by sure steps over this terrain so new and so rich, but I do not know if the zeal of one man alone and his entire life can be sufficient for such a vast enterprise. Whatever may happen, I will continue my

research and chase after the original monuments, the ONLY
GUIDES that we can follow without the risk of being held
back, as I have been *for ten years*, by the inexact inscriptions
engraved in the great work of the *Commission of Egypt.*

From this time on Champollion was not just obsessed, he was
driven. As yet he was the only person who could read the texts – he
had to translate them to make available the information they contained, but he also had to teach others his system of decipherment
and above all to publish his latest methods and results to make them
available as widely as possible.

After his initial excitement at what the collection held, he immediately started a systematic examination of the texts. For once his
health was improving and the journey over the mountains seemed
to have cured his tinnitus (the 'instrumental concert' as he called it)
that had been making his life a misery. He was also encouraged by
the warm welcome he received in Turin and the facilities that were
provided for him, including many volunteers who helped make
casts and imprints of the inscriptions and copies of texts. The stone
stelae alone, although only a small part of the collection, provided
a massive amount of information. Among them was the damaged
bilingual stone of which Young had tried so hard to obtain a copy.
The text of the stone confirmed what Champollion had already
deduced, that Caesarion, Cleopatra's son by Julius Caesar, had ruled
Egypt as co-regent with her. The parallel texts in Greek, demotic
and a few surviving hieroglyphs were useful for confirming his
method of decipherment, but against the other stelae the information it offered was of little interest – it would certainly not have
provided Young with his longed-for key to decipherment.

At this stage the Rosetta Stone inscriptions were of little further
use to Champollion – he never even published a translation of
them, and it was decades later before anyone else took the trouble
to do so. It was the fact that the Rosetta Stone had three parallel
texts, one of which was hieroglyphs, that had made it so important,

as it appeared to offer a means of decipherment and so acted as a focus of attention, stimulating new research into hieroglyphs. In reality the texts of the Rosetta Stone were of limited use because the hieroglyphs were so damaged, as Champollion had pointed out in his '*Lettre à M. Dacier*': 'The hieroglyphic text of the Rosetta inscription, which would have lent itself so happily to this research, because of its breaks only presented the single name of Ptolemy.' Having become the focus of attention for would-be decipherers, the Rosetta Stone still remains a powerful popular symbol, even though its inscriptions failed to fulfil the hopes and expectations they aroused. Other inscriptions and papyri were far more important in providing clues to decipherment.

When he turned to the pieces of Egyptian art in the Drovetti collection and began to decipher the inscriptions, Champollion found before him the actual images and names of over thirty pharaohs. Now that he could literally put names to the faces, he made the discovery that Egyptian art was not purely formulaic as had been thought. Although some statues presented pharaohs in a stylized way, others were obviously attempts at realistic portrayals – he suddenly understood that with some statues he was face to face with people who had ruled Egypt thousands of years ago: even the short inscriptions on statues were providing unimaginable results, proving that as historical documents, pieces of Egyptian art were far more useful than anything produced by the Greeks or Romans. A little later, still enchanted by the seemingly magical collection, he wrote to his brother-in-law:

Imagine if I was master of my hours, or rather of my minutes, in the middle of more than 50 Egyptian statues loaded with historic inscriptions, of more than 200 manuscripts in hieroglyphs, of 25 to 30 mummies and of 4,000 or 5,000 little figures or statuettes, nearly all carrying an inscription where I can find nectar to gather. The first excitement has not yet passed, although my whole days have been used since 12

June in studying the so curious remains of my poor old
Egypt...Nearly my entire life passes thus *in the midst of the dead
and in stirring the old dust from history.*

Absolutely amazed at the amount of information the relatively
brief inscriptions on Egyptian art could provide, Champollion soon
moved on to examine the immense numbers of papyri which had
longer texts written in hieratic and hieroglyphs, some with different
texts on both sides and even notes jotted in the margins. Papyrus,
known to the Egyptians as ⟨glyphs⟩ (*shefedw* – papyrus roll), was a
form of paper made from the papyrus plant that once grew abun-
dantly in the still, shallow waters of the Egyptian marshlands. The
plant was used to construct a variety of objects from sandals and
baskets to river boats, but for the manufacture of paper the stem of
the plant was used. After harvesting, the stems were cut to the
required length and the outer skin or rind was peeled off. The core
of the stem was then cut or peeled into strips which were laid side by
side and another layer of strips placed on top, at right angles to the
first layer. The two layers were then pressed or beaten together, and
the natural adhesives in the plant welded the strips together as they
dried out. The result was a sheet of papyrus, slightly thicker than
modern writing paper, that was ready for the scribe to use.

For many purposes a scribe would use a very basic set of equipment called ⟨glyphs⟩ (*menhed* – scribe's palette). Pens were made
from reeds, the ends of which were frayed, forming a brush-like
writing implement rather than a nib. Black and red inks were used
for general writing purposes, but other colours were used for illustrations within the text. Inks were made into solid cakes which were
held in hollows in a long rectangular palette, usually made of wood.
The scribe did not use the inks as a liquid, but dipped the reed pen
in water and then rubbed it on a cake of ink. For convenience,
sheets of papyrus were rolled and could be tied and sealed – the rolls
were usually stored in wooden chests or in pottery jars, and a long
text such as a story that was written over several rolls of papyrus

was likely to have its own chest or jar for storage.

As Champollion began to examine the papyri, it became clear that many of them were collections of spells designed to ensure the survival of a dead person in the afterlife (spells now known as 'The Book of the Dead'), but among them was a damaged plan of a tomb. Ancient Egyptian plans of tombs are rare, and this one remains the most detailed plan of a royal tomb to have survived, drawn at a scale of 1:28 with some of the dimensions labelled. By comparison with illustrations of the Valley of the Kings in the *Description de l'Égypte*, Champollion concluded that it was a plan of the tomb of Ramesses III, although it was later shown to be that of Ramesses IV.

After many days of intense and methodical work on the papyri at Turin, sometimes patiently piecing together fragments before he could read them, Champollion learned that there were some smaller fragments of papyri in another part of the museum that had not been shown to him because they had been deemed 'unusable.' He recorded his shock on seeing these fragments in a letter to his brother in early November:

> On entering this room which I will henceforth call the *Mausoleum of History*, I was seized by a mortal chill in seeing a table ten feet in length covered in its entire expanse with a bed of debris of papyri at least half a foot deep. *Quis talia fando temperet a lacrymis!* [Who could hold back their tears on speaking of such things!] In order to master my upset a little, I supposed first of all that I only saw there four or five hundred funerary manuscripts, and I had the courage to glance at the most extensive and the least shapeless: my wound reopened when I saw in my hand a fragment of a piece dating from year 24 of the Pharaoh Amenophis-Memnon. From that moment, the resolution was taken to examine one by one the large and small fragments which covered this table of desolation.

That the papyri had come to be in such a tragic state was largely due to the treatment they had suffered during transportation. In the climate of Egypt, papyrus is extremely durable – the earliest dated papyrus is nearly 5,000 years old – but inadequate care and a humid climate can quickly reduce fragile papyri to fragments. Another problem lies in the unrolling of brittle papyri, many of which languish in museums unopened, their contents a mystery, awaiting the development of new techniques of unrolling.

By the time he started work on sifting through the fragments of 'unusable' papyri, Champollion had already been in Turin for five months, working relentlessly without a break from morning to evening, and the strain was beginning to affect his health once again. With the onset of winter and the humidity caused by persistent floods in Turin during the autumn, he was suffering from rheumatism, bouts of fever and dizzy spells, a situation made worse by the increasing numbers of visitors to the museum whom he felt he could not turn away, requiring him to leave off his work and explain various aspects of the collection. More slowly than he would have wished, he reassembled many of the pieces of papyrus and began to make sense of them. What they contained filled him with awe:

> I have seen roll in my hand the names of years whose history was totally forgotten; names of gods who have not had altars for fifteen centuries; and I have gathered, while scarcely breathing for fear of reducing them to powder, such little pieces of papyrus, the last and only refuge of the memory of a king who in his lifetime perhaps found himself cramped in the immense palace of Karnak!

Then, among these battered remains of thousands of documents, he found a few tantalizing scraps that caused him great excitement and after eight days of searching, he ended up with a total of fifty pieces of a manuscript he dubbed the 'royal canon.'

The Turin Royal Canon, as it is now known, is a fragmentary

papyrus dating to the reign of Ramesses II (1279–1213 B.C.) which listed the names of previous Egyptian rulers. When Drovetti first acquired it, the papyrus was apparently largely complete and listed about 300 Egyptian rulers, but on the journey from Egypt to Italy it disintegrated and some pieces were lost. Not only did the papyrus give the names of foreign rulers, usually omitted from other lists, but it also recorded the precise duration of each ruler's reign. Champollion recognized that the list of kings was absolutely invaluable for the light it could shed on the early history of Egypt and for the establishment of a chronology, but despite a thorough search parts of the papyrus could not be found, causing gaps in the list. He felt every gap as a personal tragedy, recording his devastation in a letter to Jacques-Joseph: 'I confess that the greatest disappointment of my literary life is to have discovered this manuscript in such a desperate state. I will never console myself – it is a wound which will bleed for a long time.'

At Turin, Champollion wrote almost daily letters to his brother and friends, and he also wrote two very extensive letters on his discoveries to his patron, the Duke of Blacas, the first of which was on Egyptian art and was published in Paris in July 1824. Intended as a series of letters, in the end only one more was published, two years later. While he continued to work his way through the papyri, becoming very expert in reading hieratic (the handwriting version of hieroglyphs used in the majority of these texts), news arrived that another large collection of papyri had arrived at the quarantine station at Livorno. It belonged to the British Consul in Egypt, Henry Salt, and was for sale, but Champollion despaired of the French government's attitude to purchasing such a collection, writing to his brother that as far as he could see, 'the Egyptian monuments will abound everywhere except in France...there will soon be an Egyptian museum in the capital of the Republic of San Marino whilst we will only have at Paris isolated and dispersed pieces.' Other news of a far more unsettling kind had also reached him, for his new patron King Louis XVIII had died on 16 September, and the Count of Artois had succeeded him as King Charles X. More reactionary than his brother Louis XVIII, the Count of Artois had been the focus of support from the extreme Royalist Ultras who had caused so much trouble for Champollion in the past. However, as Champollion had been reconciled with Louis, who had financed his trip to Italy, it was possible that he could establish a similarly good relationship with the new King, particularly since Charles was known to be more interested in scientific projects and less parsimonious than his brother Louis. Champollion decided to try to remain optimistic.

Champollion also faced some difficulties in Turin, because of a growing enmity between him and the director of the museum, Cordeo di San Quintino. Friction between the two was probably inevitable since Champollion had been hailed as a celebrity and an expert, and for months had virtually taken over the museum which was normally San Quintino's own personal province. Passionate about the objects he was studying, Champollion was free with his advice and not always tactful in the way he gave it, while San Quintino feared for his position as museum director and became resentful at interference from a foreigner who had powerful friends.

Although there was still more to be done, by January 1825 Champollion felt he should move on to other places in Italy, hoping to return to Turin later to complete his work. San Quintino and his supporters saw this as an opportunity to try to organize the Drovetti collection at the museum and write the catalogue in order to remove the reason for Champollion's return. This spurred San Quintino's opponents into pushing forward Champollion's membership of the Academy of Turin, and he and Jacques-Joseph were unanimously elected in mid-January. Champollion finally left Turin for Rome at the beginning of March, making the journey in ten days with brief stops on the way at various places such as Milan and Bologna, where he managed to make a rapid examination of other Egyptian collections and copy their hieroglyphic texts. Having travelled most of the time through torrential rain, with sleepless nights in uncomfortable inns, he arrived in Rome at six in the morning on

11 March, exhausted and with his hands and feet swollen. Even so, he could not contain his excitement and, like any tourist in Rome for the first time, he barely took time to check in to his hotel before he was out again to see the city, as he admitted to Jacques-Joseph: 'I went right to St. Peter's: seeing that my appetite was at its height, it was necessary to begin with the choicest morsels. To describe the impression that I experienced in arriving at the site of this basilica is an impossible thing: we are wretches in France; our monuments are pitiful by the side of these Roman splendours.' He continued on a rapid tour of the city, particularly noting the obelisks and other Egyptian monuments that he saw and did not arrive back at his hotel until after midnight. Having described the exhilaration of the day in a letter to his brother, he ended with: 'Here is my first day in Rome. I will never forget it!'

Despite the first good night's rest since he had left Turin, his feet were still swollen the next morning and he could hardly walk, but a friend put a carriage at his disposal. Champollion spent the next four days visiting friends and contacts in the city and seeing as many of the sights of Rome as he could manage, but aware that both his time and money were limited he soon continued on his way south to meet Blacas, who by this time had taken up his post as French Ambassador at Naples. After all the travelling and his short but frenetic stay in Rome, Champollion was in great need of rest, and although he had to fulfil various social commitments, he was for once largely free from the continual examination of Egyptian artifacts and texts that had become his daily life in Turin. His main work was to prepare a report on the Canopic jars and other Egyptian objects in the possession of the new King of Naples, Francis I, and he had several audiences with the Queen, Isabella, who required him to explain his system of deciphering hieroglyphs. This was far from being the first time he had been asked to explain his methods, and as his ever-increasing fame spread, more and more of the people he met clamoured to hear of hieroglyphs from the mouth of the man who had mastered their secrets.

Determined to make the most of his brief stay in Naples, Champollion visited the ruins of Pompeii which were being disinterred from the ash and lava that had buried the Roman town when the volcano Vesuvius erupted in A.D. 79; the last serious eruption had occurred only thirty years before his visit, missing Pompeii but with lava flowing out to sea. He recorded his amazement at the newly discovered ruins in a letter to Jacques-Joseph:

> the day of the 1st April passed like a minute. It would require a volume to give an exact idea of everything that one saw there. I explored the market place, rushed to the *Forum*, said my paternosters in the temples of Mercury, Neptune, Jupiter, Diana and Venus; then, a long meditation in the temple of Isis, and, mixing the profane with sacred, I went to the two theatres that I quickly left in order to reach the amphitheatre in time...finally I rushed through *the streets*, entering a crowd of houses where more or less curious frescoes are found. I especially admired two paintings still in place, discovered about a month ago...the drawing was admirable and the colours were excellent. These are the most beautiful ancient paintings known, in my opinion at least.

He had yet to see the tomb paintings of Egypt.

Although impressed by Pompeii, what Champollion really wanted to see was the ancient Greek and Roman town of Paestum, also called Poseidonia, with its three beautifully preserved early Greek temples, but everyone tried to dissuade him, since it was in the middle of an isolated and unhealthy marshland that was notorious for brigands. Despite the warnings, stubbornly he set out from Naples on 10 April and slept that night at the town of Eboli. The next day he was deeply moved by the spectacular ruins:

> after three and a half hours of walking, thanks to my carter who lost the good road in order to keep to the foot of the

mountains, in the middle of rocks, I finally caught sight of the ruins of old Poseidonia, scattered in a desolate plain. The ruins of Paestum have been described a hundred times. Nothing is more simple than the architecture of its buildings, but it is impossible to give an account of the effect and to convey the profound impression that one experiences at the sight of three Greek temples in an astonishing state of preservation and which without any doubt date back to the most ancient time of prosperity of the Greek colonies in Italy...At a certain distance and especially because they stand out as golden yellow against the beautiful azure of the sky and the sea, I believed I saw *Egyptian temples*...I was already in love with the old style, but it is now a declared passion. It is useless to add that, like all those who have visited Paestum (and the number is not very great because of the new-style Greek heroes who often infest the countryside), I maintain that there is nothing more beautiful and imposing in Italy, and emphasize that I even include Rome when I pronounce this sentence...There is no other noise in this enclosure than the cries of crows or buffalo, which seem to give a marked preference to the beautiful temple of Neptune. Some crows flit about in its forests of capitals, or perch on the cornices, while others rest in the shadow of the robust columns of the peristyle. I will never forget such a picture, and of all my excursions it is that from which I will keep the deepest memory.

He slept that night at Salerno and the next day returned to Naples, stopping on the way to spend another four hours at Pompeii where he 'made a libation of lachryma-christi in the temple of Isis...and a second in that of Venus.' Using a local wine, lachryma Christi, that derived its name from the legend that Christ, looking down on the beautiful Bay of Naples, wept for the sins of its inhabitants, he had made offerings in the temple of the Roman goddess of love and the temple of Isis, the chief goddess of Egypt. The symbolism of these acts can be interpreted in many ways, perhaps even as a willingness to sacrifice his Catholic faith to love and to Egypt, but Champollion left no explanation of these libations and, outwardly at least, remained a Catholic for the rest of his days, leaving a question mark over the true nature of his beliefs.

Now in much better spirits than when he arrived in Naples, he finished the letter to Jacques-Joseph in which he described Paestum with: 'My health is perfect, except for the usual pains. Look after your teeth, of which we have the most pressing need in order to defend ourselves: mine are in a good state and only ask for something to bite.' Having visited those sites that he most longed to see, Champollion felt he could not allow himself more time in Naples when he still had so much work to do, and so ten days later he was back in Rome, where he immediately began a study of the Egyptian obelisks which had been transported to the city by various Roman emperors over 1,500 years before. As he undertook the difficult work of copying the inscriptions on the obelisks amid the unsavoury ruins that often surrounded them, baked and basted by the alternating sunshine and showers of April in Rome, he came across more and more errors in the previous copies of the inscriptions. The versions of the inscriptions published by Kircher, the seventeenth-century priest who wrongly believed he had managed to decipher hieroglyphs, were particularly inaccurate, and Champollion was strengthened in his opinion that he must see as many original inscriptions as possible: better than Italy, the obvious place to go was Egypt itself.

At Rome he initially lodged with Blacas, who had journeyed back there with him and introduced him to many influential people, but this was a mixed blessing for Champollion, much of whose precious time was taken up in meeting people who wanted a first-hand explanation of the decipherment. Finding himself once more a celebrity, at a time when he would have preferred to be a secluded scholar, he did at least benefit from the scramble to offer him accommodation when Blacas left Rome. As well as the public monuments such as the obelisks, he now began to study all the public and private collections of Egyptian material to which he could gain access, including the collections of the Vatican, some of which he had already seen and studied when a student in Paris – part of the loot from the Vatican that had been returned to Rome after the final exile of Napoleon.

On 15 June 1825, just two days before he was due to leave Rome, Champollion was granted the honour of being presented to Pope Leo XII, who received him warmly, and he afterwards wrote to his brother that 'the Pope, who spoke French very well, was pleased to say to me three times that I had *rendered a beautiful great and good duty to Religion* through my discoveries.' The Pope's enthusiasm was based more on Champollion's dating of the Dendera zodiac, which had temporarily silenced opponents of the Biblical chronology, than on the decipherment of hieroglyphs, and in his enthusiasm the Pope offered him an appointment as a cardinal. Amazed and embarrassed, he excused himself on the grounds that he was married and had a daughter. Feeling that some honour was nevertheless due to Champollion, the Pope used his influence on the King of France and just over a month later the French government informed Champollion he was to be made a Knight of the Legion of Honour – the same distinction that Jacques-Joseph had received from Napoleon ten years earlier.

As Champollion had already come to realize, the favour and friendship of people in power all too often carried the penalty of arousing the jealousy and hatred of those lacking such influence, and so recognition from the Pope brought a new chorus of voices willing to speak against him, while not silencing all opposition from Catholic churchmen, some of whom still considered his work a threat to Church authority. Among scholars across Europe a shifting pattern of opposition and support was emerging. Some, like San Quintino, the director of the museum at Turin, were motivated to oppose Champollion and ally themselves with his rival Young, while actually using Champollion's system of decipherment simply because it worked. Others, realizing their error, simply transferred their support away from Young, including Sir William Gell, who had by now met Champollion in Rome and was convinced of the truth of his system. While many German scholars supported Champollion, a few were becoming a nuisance. Friedrich-August Spohn, eighteen months younger than Champollion and Professor of Greek and Roman Literature at Leipzig, claimed to have deciphered hieroglyphs, but he died in early 1824 before his theories could be put to the test. The work of completing his publications was entrusted to the German Orientalist, Gustavus Seyffarth, who embarked on a lifetime of opposition to Champollion. Meanwhile, Heinrich Julius Klaproth, an accomplished German Orientalist who had settled in Paris in 1815 but with the title and salary of Professor of Asiatic Languages at Berlin, was now totally opposed to Champollion, having previously supported him. From the time of Champollion's departure for Italy, when he finally gave an unequivocal refusal to become involved in Klaproth's academic intrigues, this scholar began a vendetta against all Champollion's discoveries, often in anonymous pamphlets, that continued to plague him for years to come.

From Rome Champollion moved to Florence in the Grand Duchy of Tuscany, and although he was there for little more than two weeks, that was long enough for him to fall in love with it. He saw it as 'the *only town* of Italy where one may enjoy a true and just freedom; it is in fact the only country where there is a government – and that is quite something.' Once again he fed on the intoxicating mix of social and scholarly engagements that was almost becoming routine, trying to squeeze in the invitations to meet influential people and explain his discoveries without curtailing his study of the collections of Egyptian antiquities, and he was invited by the Grand Duke Leopold II to catalogue his recently formed collection. What Champollion found most refreshing, particularly after his experiences in Rome, was that in the climate of intellectual freedom prevalent in Florence, his work was judged as he wanted it to be judged – from the viewpoint of logical and scientific enquiry, without bias and religious bigotry.

Leaving Florence in early July, Champollion's last stop was to be Turin before returning to France, but he made a rapid detour to Livorno to look at the collection of Egyptian antiquities and papyri that he had heard about some months earlier and which was up for sale. The name of the owner of the collection was supposed to be secret, but he knew it belonged to Henry Salt, and when he actually saw what was in the collection, he was galvanized into writing to Jacques-Joseph: 'This collection, more beautiful than that of Drovetti (if one excludes the great statues which are missing), is for sale and the government would be able to have it for 250,000 francs.' Champollion contacted Blacas with details of the collection in the hope that this time the French government would act, and at Turin he soon wrote again to his brother, emphasizing that Salt wanted to sell to France: 'I have said to you, and I repeat it, that it is being given away for nothing, and the honour of France is concerned in not letting the fruit of the long labour of a Goddam [derogatory term for an Englishman] escape who, although through his own national pride does not want to be named, is somewhat inclined to make money from French Frogs.'

Henry Salt, then forty-five years of age, had been in Egypt as British Consul for nearly ten years. His first collection of Egyptian antiquities had been sold to the British Museum three years earlier, but he was unhappy with the transactions over its purchase. Contrary to Champollion's view, his wish was not to sell his second collection to the French but to the English, as he had written to a friend: 'It would be a great pleasure to one that it should go to England; but no more of *dealing* with the British Museum.' What Salt desired most of all was to retire from Egypt and be awarded a pension: 'I have collected, and my collection is now in Leghorn, antiquities to the value of four thousand pounds: the finest collection of papyri existing, the best assortment of Egyptian bronzes, several paintings in encaustic, and rich in articles of gold and porcelain – in fine, what would make the collection at the [British] Museum *the choicest in the world.*'

Back in Turin, as Champollion struggled to finish cataloguing the collections there, he gradually despaired at the inaction of the French government regarding Salt's antiquities, but at least the museum director, San Quintino, left him alone to continue his work. He was also encouraged by a book on hieroglyphs (*Essay on Dr. Young's and M. Champollion's Phonetic System of Hieroglyphs*) that had just been published by Salt, a copy of which arrived from the author himself. Through supplying antiquities to collectors in England, Salt had become a supporter of Young's claims, but his book revealed that he too now accepted the validity of Champollion's discoveries. Only the year before he had written as much to William Hamilton: 'You will be surprised to hear that I have become a complete convert to Mons. Champollion fils' [*sic*] system of explaining the hieroglyphics...I soon began to find that I had been wrong in ridiculing it.' Salt's own attempts at decipherment hardly qualified him as a serious rival though, and his book has been described as 'an unhappy work, in which whatever was his own was hopelessly and utterly wrong.' Champollion's spirits were also lifted when he found out that he himself was being hailed as a celebrity in several foreign newspapers, because various diplomats at Rome had sent their governments favourable reports on his work, which had filtered down to some journalists – he hoped that this unexpected fame would lend weight to his lobbying of the French government to purchase the Salt collection.

It took Champollion nearly three and a half months to finish his work at Turin, by which time it was early November and there was a danger that the Alpine passes would soon be closed. Crossing the Mount Cenis Pass, he was delayed by a terrible storm that caused havoc in the surrounding villages, and he arrived in Grenoble at ten o'clock in the evening, surprising his wife who had not thought to see him before the next day. After a year and a half, he was amazed at how his daughter Zoraïde had grown, but the long absence had not diminished his adoration of her: 'If I didn't have the honour of being her father, I would say to you that her reputation which already extends into two or three districts, is perfectly deserved – that she is the prettiest little child in the Dauphiné.'

Champollion was also reunited with Jacques-Joseph, and the two brothers had much to discuss that could not be said even in the frequent and often frank letters that they had exchanged, including opposition in Paris to the French government purchasing the Salt collection. Despite all the praise and honour heaped on Champollion by the Pope, much of the continuing opposition came from Catholic churchmen and arose from his studies in Turin. While Champollion had preserved the Biblical chronology with his redating of the Dendera zodiac, he had equally endangered it by the discovery of the Royal Canon, which mentioned dynasties of very early kings dating back far beyond the accepted beginning of the world as envisaged in the Biblical chronology. Because it was in such a fragmentary state, more like a jigsaw puzzle than a manuscript, the Church could easily dismiss the Royal Canon, but each new collection of papyri from Egypt might contain proof that the Church was wrong in its interpretation of the Bible, and many churchmen did not want such proof endorsed by purchase of the Salt collection. For the time being Champollion kept to himself this proof that the Egyptian dynasties were so early in date because the evidence was not yet overwhelming or indisputable. Nevertheless his enemies in the Church rightly suspected the true situation and maintained their efforts to discredit him.

Jacques-Joseph soon needed to return to Paris, but Champollion remained at Grenoble, continuing to make revisions to some of his publications while waiting to hear the decision about the Salt collection. A few months earlier, in July, a rival collection of antiquities belonging to a former Italian horse-dealer turned excavator and collector, Giuseppe Passalacqua, had arrived in Paris. Edme Jomard, editor of the *Description de l'Égypte*, was lobbying hard for the government to buy it, because he was most likely to be given the

post of curator to look after it, whereas if the Salt collection was pur-
chased, Champollion would probably be given the post. Since
Champollion had been offered several posts abroad, his enemies
reasoned that being blocked from the post of curator in Paris would
force him out of the country, and so they rallied behind Jomard.
Although the Passalacqua collection was much cheaper than the
Salt collection, the King wavered, and controversy raged in the
Paris newspapers.

The struggle continued into the early months of 1826, with
Jacques-Joseph in the thick of it in Paris. At one point it looked as if
the matter was settled in favour of Champollion, which prompted
open denunciations of the 'injustice' of such a decision from Jomard
and his supporters, but towards the end of February Champollion
was informed that the King had allocated 5,000 francs for him to go
back to Livorno to study the Salt collection and give his opinion on
the price – it looked as if he might have won the contest. Seizing the
opportunity, he immediately made preparations for the journey
and left Grenoble on 1 March. It was too early in the season for the
passes to be open, and the Mount Cenis Pass was still snowbound,
as he described in a deliberately understated letter to his brother:

> I got back into the carriage and only got out of it ... at the foot
> of Mount Cenis, in order to dive into a chest four and a half
> feet in height, set on a sledge attached to the tail end of a horse
> used to running on snow. The climb was done rather peace-
> fully, as was the descent, but not without a certain terror: it is
> difficult to slide on a road covered with twenty to thirty feet
> of snow – the road I had crossed on land four months
> previously – without experiencing a certain emotion.

He was only just in time, as his first letter from the Italian side of the
Alps recorded: 'in three days it will only be possible to cross at the
risk of one's life. The snows are beginning to melt, and already two
avalanches have bitten into the road and buried some of the valleys.'

Arriving at Livorno in mid-March, Champollion started work immediately. The collection had increased in size, because when he had first seen it the larger pieces of sculpture and a huge stone sarcophagus were still on their way from Egypt to Italy. Now he had to check painstakingly through each of the items of the collection – nearly 5,000 pieces in all. News soon reached him that the report on the collection by the Duke of Blacas had now been accepted by King Charles X and the purchase was authorized, so his task changed from one of straightforward assessment to ensuring that everything could be accounted for so that nothing would be lost in transit, and he eagerly began preparations for the packing and transportation of the antiquities to Paris.

Champollion's fame was such that almost everywhere he went academic institutions welcomed him, honoured to have him as one of their members, and at Livorno he was made a member of their Academy of Sciences. Here he was introduced to Ippolito Rosellini, Professor of Oriental Languages at Pisa, who wanted to become his pupil and had obtained a grant to cover his expenses if he should need to travel to accompany Champollion. Twenty-five years old, Rosellini had already spent a year studying hieroglyphs, and Champollion accepted him as a student without hesitation. Membership in this Academy also led to another fateful meeting: on 2 April it held a public meeting to honour him, and one of the speakers was Angelica Palli. This twenty-eight-year-old woman was the daughter of one of the foremost businessmen in Livorno and already had a growing reputation as a poet. At the meeting she recited a poem in honour of Champollion, who was touched by this gift of words and considered it 'the sweetest reward that I have received for the Egyptian dust being my staple diet for fifteen years' – he soon fell in love with her. To his friend Costanzo Gazzera, a librarian at Turin University who was himself becoming proficient in hieroglyphic decipherment using the system laid down by Champollion, he described her as:

rather educated, of a perfect kindness, and you must without doubt have heard talk of her if you are not barbarians. As for me, I thank the great Amun-Ra for having made her acquaintance and for having found favour in her eyes; but remembering that the mummies also have rights, though silent, I remain in the midst of them as much as possible, and I only see the amiable Sibyl [Angelica Palli] occasionally, for fear that Hathor [Egyptian goddess of love] might get mixed a little too much in my gratitude.

The handful of letters from Champollion to Angelica Palli that have survived make it clear that although he was in love with her, friendship but not love was reciprocated. She subsequently married a son of one of the oldest families in Livorno, but for a few brief weeks she dominated his thoughts as he prepared the Salt collection for transportation and awaited the arrival of a ship to take it to France. By the end of April he had written his report on the collection, which had been published in Paris in an attempt to silence his enemies who were crying out against the King's purchase, but by June he was still waiting for the ship. Fretting over the wasted time, he felt he could not return to France without seeing everything safely stowed aboard ship, especially since a letter from Jacques-Joseph had arrived informing him that in mid-May Champollion had finally been appointed to the coveted post of curator in charge of a new Egypt section of the Louvre Museum – the Salt collection was now fully his responsibility and at long last he had acquired a professional post after so many years in the academic wilderness.

For someone used to working almost every minute of the waking day, the weeks he was forced to spend at Livorno waiting for the ship to arrive were a total loss to Champollion, but throughout this time Rosellini was with him, gleaning as much as he could about hieroglyphs and Egyptian artifacts – knowledge which in time Rosellini and his own students would use as the foundation for the study of Egyptology in Italy. The ship finally arrived in late June, and by the second week in July the collection had been loaded and was on its way to the port of Le Havre at the mouth of the River Seine in northern France, and then by river to Paris. A few weeks remained before Champollion was needed back in Paris to unpack the Salt collection and he was determined to make the most of them. Travelling via Pisa and Florence, he and Rosellini arrived in Rome in mid-July.

Champollion resumed recording the inscriptions on the Egyptian obelisks in Rome and correcting the proofs of the engravings of his earlier drawings. He was working towards a complete and accurate record of all the inscriptions on these obelisks, supported by a promise from the Pope of funding for publication. He spent about three weeks in Rome, during which time a rare event took place – a direct confrontation between Champollion and one of his rivals, the German Orientalist Gustavus Seyffarth. With other rivals, such as Jomard, Roulhac and Young, personal meetings were not confrontational but regulated by the rules of social etiquette. Seyffarth was different. He was travelling around Europe studying Egyptian collections and Coptic manuscripts in order to complete the late Spohn's publications, in Latin, on his alleged decipherment of hieroglyphs. There is no doubt that Seyffarth was a brilliant scholar, probably equal to Champollion, but he clung to some fantastic notions about certain subjects such as hieroglyphs, and utterly wasted his genius. Denying most of Champollion's work he promoted the idea that hieroglyphs were based on an earlier form of Coptic, which in turn was derived from Hebrew, and that all hieroglyphs represented syllables and originated from Noah's alphabet. Hearing that Champollion was in Italy, Seyffarth deliberately went to Rome to challenge him before expert witnesses so that they could decide whose decipherment system was correct. Champollion rose to the occasion, declaring 'I will slay him at the foot of the Capitol.'

In a letter to Thomas Young in early August, Sir William Gell reported on his own meeting with Seyffarth and on the actual confrontation that subsequently took place between Seyffarth and Champollion: 'I am not aware whether you have any communica-

tion with Champollion, but we have just had him in full swing at
Rome, where he was preceded about ten days by Mr. Professor Seyffarth, his antagonist. Seyffarth is a very nice gentlemanly sort of
a Lord Palmerston.' When Gell tested Seyffarth on deciphering
names in hieroglyphs he was only correct on those names already
deciphered by Champollion,

> but when I tried him with those he had never seen or heard of
> fresh from Egypt, he could do nothing, though he had his
> great quarto book full of plates in his hand...Well, I had settled
> in my own mind that the said Mr. Sighpoop [Gell's derogatory
> name for Seyffarth] was crazy, when Champollion came [to
> Rome], and I proposed that they should fight with two obeliscs
> for swords, and the labrum [a massive granite basin] of Monte
> Cavallo for a shield. They met at Italinsky's and Nibby, who
> owned the influence of the great quarto very powerfully, says
> Italinsky was on the side of Sighpoop, but all the spectators
> say the contrary, and I met them next day at the French min-
> ister's. Champollion asked him into what language he
> translated his hieroglyphics, to which he said 'Coptic.' Then
> says Champollion, 'I will not say there is no sentence, but
> there is no word of Coptic in your translation.' 'Oh,' says Seyf-
> farth, 'it is a more ancient Coptic than that of the books.' C.
> 'Where did you learn it?' S. 'In the Rosetta inscription.' C. 'In
> the two lines you have published?' S. 'Yes.' C. 'Then give me
> leave to say that as you have published them they are so falsely
> copied that they give no idea of the real figures, and that no
> ten figures together are correct. Moreover, the name you have
> chosen is the one ill written in the original ... ' (all which is
> true). The fact is, Seyffarth knows nothing of the monuments
> and never saw any. He answered nothing, and told Nibby he
> thought it was better to be silent, as Champollion was so vio-
> lent, which I understand he was not. All I can say is, that even
> the Germans did not support him, and that as all his figures

may mean any letter of the alpbabet [*sic*] according to their position, if his scheme be true it cannot be worth learning...I am quite astonished to find how Champollion has *progressed*, as the Americans say, this year, in applying the Coptic words to the figures.

Champollion's own observation was that this rival had been vanquished: 'I have shown him, bluntly, all the wrong of his business and put forward arguments to which he has not known how to reply. The floor has judged his silence. I see him every day now, but there is no longer any talk of hieroglyphs...he is a man ruined in Italy.' Many scholars thought that Seyffarth's system was bizarre and unproven, and opposition mounted. Later that year Champollion would write from Paris to Rosellini: 'If you find Seyffarth at Turin, preach to him again so that he converts and stops making himself ridiculous by his absurd dreams. He is held up to ridicule in Germany, nobody takes his side in France. You know what his position is in Italy – it will be a pious thing to do. Don't fail if you can.' Back in Germany in 1828, Seyffarth found his situation there increasingly difficult and emigrated to the United States some twenty-five years later, where he continued to campaign vigorously against Champollion and his followers.

After the confrontation with Seyffarth, Gell was a fervent supporter of Champollion, much to Young's chagrin. Still working at St. George's Hospital, Young now moved to a spacious and elegant house at 9 Park Square in London, where he continued to correspond with scholars across Europe on the subject of hieroglyphs. His main concerns were twofold: the study of demotic and his ever-persistent efforts to persuade scholars that he was first to decipher hieroglyphs. It cannot have helped his bitterness that Gell continued to praise Champollion highly in his letters, commenting on his willingness to share information: 'I beg to state that so far from hiding his new discoveries, the said Champollion has given me so many things not published, that, if I were inclined, I could pretend

I were the inventor of as much again as he is.'

In Rome, Champollion approached those foreign diplomats who had been so impressed by him during his previous visit – sounding them out about the possibility of setting up a European expedition to Egypt under his direction with experts drawn from several countries. Convinced that his next step must be to visit Egypt to study the inscriptions and monuments there, he had already approached Blacas on the subject and had obtained his support. In his mind Champollion was mapping out the immediate future: he would use what time was left before the Salt collection arrived in Paris to achieve as much as possible in Italy, then he would spend about a year setting up the new section of the Louvre Museum, original ideas for which he was already forming, and at the same time he would assemble his expedition to Egypt. He began to canvass potential expedition members as he travelled through Italy with Rosellini, who was learning so rapidly and becoming such a useful assistant that he was automatically a candidate.

Moving from Rome south to Naples, Champollion found time for another visit to Paestum, where the temples had made such an impression on him. In Naples he also met up with Gell, and taking Gell's recently published book, *Pompeiana*, they made a tour of Pompeii. Impressed by his work, Champollion enlisted him as another potential member of his Egypt expedition. Blacas meanwhile was conducting excavations of the Etruscan and Roman town of Nola, where the Roman Emperor Augustus, conqueror of Egypt, had spent his last days. Champollion visited the site, making the most of the pure air of the Naples area which he found restored his health and strength better than anything else. He had intended to return to Rome to complete his work, but by now the plague was raging there and anyone who could, including the Pope, had deserted the stricken city. As time was running out and he would soon be needed in Paris, he travelled north through Livorno and Florence with a detour to Venice, arriving back in Grenoble in late October 1826. He had been sad to leave the 'sparkling blue sky' of Italy and his many friends there, including Rosellini, and he was not helped by a deterioration in the weather as he travelled north, which reversed all the benefits of his stay in Naples. As soon as he reached Grenoble Champollion was struck with a violent attack of gout in his right foot, but there was no time to rest, since both his family and that of Jacques-Joseph were just finishing their preparations to accompany him to Paris, where for the first time they would all live in the city under one roof at 19 rue Mazarine – after five days' travel, they arrived there on 20 November.

Work now began on the Egyptian section in the King Charles X Museum within the Louvre Palace, where it occupied four rooms on the upper floor of the Cour Carrée (Square Courtyard) flanking the River Seine, with some ground-floor rooms to display those sculptures too heavy for the upper floor. Champollion wrote to Gazzera in Turin at this time describing his new museum: 'I have a magnificent room on the ground floor for my large pieces, and four rooms on the first floor of the Palace. So, here will I soon be in the midst of painters, architects and masons, and if things proceed it will not be without difficulty.' The difficulties immediately arose as he ran into opposition, partly from old enemies and partly from new ones. Jomard in particular was very bitter that Champollion had been given the position of curator and continued to intrigue against him, but much of the opposition came from people directly connected with the museum who regarded him as an interloper and did not like his ideas on how the new Egyptian section should be arranged. In a letter to Rosellini, Champollion complained: 'My life has become a struggle…My arrival at the museum is upsetting everyone, and all my colleagues conspire against me because instead of considering my place a sinecure, I intend to occupy myself with my division, which will necessarily reveal that they do nothing at all with theirs. There is the nub of the affair! A battle is needed to obtain even a nail.'

He lost many of the battles: despite his vigorous protests the rooms for the display of the Egyptian collections were decorated in

Classical rather than Egyptian style, and in those ceiling paintings that had any relation to Egypt the allusions were Biblical or Classical and were not chosen by him. He did manage to have Jean-Joseph Dubois, the artist responsible for his *Panthéon* publications, appointed as his assistant, and he fought hardest to establish his own arrangement of the exhibits according to logical and scientific principles. From his experience as a teacher, Champollion wanted to use the exhibits to inform and educate visitors to the museum – a complete departure from the usual museum practice of arranging objects in relation to one another to show off their artistic merits to best advantage, and one by which he was setting new standards. He considered it necessary to arrange the exhibits:

> so as to present as completely as possible the series of divinities, the series of monuments which quote the names of the rulers of Egypt from earliest times up to the Romans, and to classify in a methodical order the objects which relate to the public and private life of the ancient Egyptians. One will therefore have in this way the systematic collection of the monuments relating to religion, to the history of the kings and to the civilian customs of the Egyptians.

Since most of the exhibits had inscriptions in hieroglyphs, Champollion had only to read them to decide on a logical system of arrangement. His colleagues in charge of other collections lacked this advantage, since there were few inscriptions on the Classical antiquities, and it would take many years of archaeological research before their collections could be arranged in such a way.

While his work at the museum was such a struggle, Champollion was at least happy at 19 rue Mazarine (not far from his former lodgings at number 28), where his own and his brother's family formed a harmonious 'Grenoble colony,' and if he failed to educate his colleagues at the Louvre, he found immense pleasure in teaching and entertaining the children at home. He had not lost sight of his plans for an expedition to Egypt, and indeed Drovetti, the French Consul there, was lobbying hard for an expedition to record and salvage as much as possible before all the monuments were destroyed. The ruler of Egypt, Mehemet Ali, was intent on the economic development of the country and had decided that increased sugar and cotton production was the way to greater prosperity. Factories to process the sugar cane and mills for cotton were being built, and the stone for their construction was being plundered from the ancient Egyptian monuments. The sooner Champollion could reach Egypt the more likely it was that the most important monuments could be identified and perhaps preserved, but through his contacts at Court, he knew that he still had enemies close to the King, and there was no hope of organizing an expedition before the new Egyptian section of the Louvre was finished. Champollion had no choice but to endure the daily battles at the museum until the task was complete.

A second collection of antiquities amassed by Drovetti was now up for sale, but even though he possessed the first collection, the King of Sardinia-Piedmont refused to purchase this new collection for Turin, so Champollion and Jomard (harmoniously for once) both worked hard to obtain it for France. Drovetti had already primed King Charles X with gifts, including that of a monolithic shrine, and by the autumn of 1827 the collection was purchased and added to the Salt collection in the Louvre. Writing to Gazzera at Turin in September, Champollion was exultant: 'A rather more important purchase has just been made: it is that of the new collection of Drovetti, which is at *Paris*...and which possesses Egyptian jewellery of an unbelievable magnificence...most of these objects carry royal inscriptions, for example a cup in solid gold.' He continued: 'This collection contains, moreover, statues, fifty Egyptian or Greek manuscripts, 500 scarabs, vases, eighty steles etc etc. We are finishing, as you see, by being more beautiful and rich than you, who could be the *Foremost* and haven't wished it.'

Champollion was now struggling against the odds to have the

Egypt exhibition ready for an official opening to mark the King's birthday on 4 November, driven as well by first-hand information about the situation in Egypt. Drovetti himself had arrived in Paris in August and reported two disturbing pieces of news – that the destruction of monuments was accelerating and that the political situation in the eastern Mediterranean was likely to deteriorate to a point where access to Egypt could become impossible. Since 1820 the Greeks had been fighting a war to establish their independence from the Turkish Ottoman Empire. The major powers in Europe had at first refrained from involvement, but in 1825 Mehemet Ali provided Egyptian aid to Turkey, prompting Russia and Britain to put pressure on Turkey to come to some sort of agreement with Greece. As the situation worsened, Britain, France and Russia signed a treaty in July 1827 pledging to use force, if necessary, to obtain an agreement from Turkey, with the result that on 20 October squadrons from the British, French and Russian fleets destroyed the Turkish and Egyptian navies at the Battle of Navarino. Nevertheless, Mehemet Ali remained favourable to the French, and a statement from him assuring protection for Europeans in Egypt was published in Paris newspapers at the end of the year.

With the redecoration of the rooms not yet finished, the opening of the Egyptian exhibition in the Louvre missed the 4 November deadline, and so King Charles X officially opened the new museum in mid-December. By that time Drovetti had left Paris to return to Alexandria, without any definite possibility of a scientific expedition to Egypt, but with both his and Salt's second collections transforming the Louvre's Egyptian section into one of prime importance. Tragically, Champollion would never meet the British Consul whose collection had done so much to transform his own life in Paris, for Henry Salt had recently died in Egypt on 29 October 1827, at the age of forty-seven years, after trying for so long to be allowed to go home to retire. Buried at Alexandria, his tomb inscription included the words: 'His ready genius explored and elucidated the Hieroglyphs and other Antiquities of this Country.'

Young, meanwhile, had finally taken the decision to give up studying hieroglyphs and to concentrate solely on demotic. Finding out that the Reverend Henry Tattam, a scholar in England, was compiling a Coptic grammar, he offered to write a demotic dictionary to accompany it, but warned Tattam at the end of the year that it would take several months to complete. Ironically, while Champollion's fortunes were in the ascendant and he had rich collections at his disposal, it was now Young's turn to complain about the lack of suitable texts to study: 'Like the dog in the manger, these people neither eat their hay nor let me nibble at it. I cannot get a single line of all the demotic contracts on papyrus which I know to exist all over Europe.'

Despite poor health brought on by the strain of finishing the exhibition, Champollion was feeling more optimistic that an expedition to Egypt could be organized, particularly as changes of ministers within the Court had removed some of his enemies from key positions, and the acclaim that greeted the newly opened museum had raised his standing. By the spring of 1828 supporters of Champollion at Court had succeeded in interesting the King in the project, and in late April he promised his support, clearing the way for Champollion to proceed – it was to be a joint French-Tuscan expedition with the protection of Charles X, King of France, and Leopold II, Grand Duke of Tuscany, under the overall command of Champollion with Rosellini as his assistant.

Having been dreaming about and planning such an expedition for over two years, Champollion enlisted Jacques-Joseph's help and lost no time in setting his plans in motion, contacting the people who were to accompany him and arranging all the stores, equipment, and diplomatic and legal documentation, so that everything was ready in under two months. In June Young came to Paris to take up his newly awarded seat as one of the eight foreign associates of the Academy of Sciences, where he was very warmly welcomed. He also had the opportunity to meet Champollion, who generously arranged to have demotic papyri copied out for him even while engulfed in preparations for his expedition to Egypt. Obviously humbled by Champollion's help and overawed by his success, Young wrote to the astronomer Arago in early July: 'I am most ready to admit, that the more I see of his researches, the more I admire his ingenuity as well as his industry; and I must be eager to bear witness on every occasion to the kindness and liberality which he has shown me in either giving or procuring for me copies of every thing that I have asked for.'

By the end of July Champollion was at Toulon, waiting for favourable winds before his ship could sail, but meanwhile the situation in Egypt had deteriorated because news of the destruction of the Egyptian navy had finally reached the Egyptian people. Alexandria was up in arms against Europeans, and Drovetti judged the situation much too volatile for safety. At the beginning of May he had written a letter to Champollion warning him not to set out for Egypt, but the letter took nearly three months to reach Paris, by which time he was already in Toulon. Jacques-Joseph opened the letter and sent a copy to Champollion, but it failed to reach him in time – on 31 July 1828, some thirty years after Napoleon and his savants had sailed from the same harbour, Champollion set out on his own expedition to Egypt.

# CHAPTER NINE

# (The Translator)

The *Églé* was a fast, well-armed corvette, normally used for escorting merchant ships between France and the eastern Mediterranean, but with the unstable situation in the East, owing to the continuing struggle for Greek independence, French trade with the eastern Mediterranean had almost ceased, and the *Églé* had been assigned instead to transporting Champollion's expedition to Egypt. Champollion had parted from his family with the optimistic words 'Don't worry, the gods of Egypt are watching over us,' and despite one bad storm, the winds were favourable and the crossing to Egypt was made in nineteen days. Without being aware of it, the expedition had easily outrun the vessels that had been dispatched to chase after it with orders to inform Champollion that the situation in Egypt was too dangerous and his expedition must be abandoned. Instead of having his hopes dashed by a forced return to France, he landed at Alexandria on 18 August 1828 and immediately touched the soil of the land that had completely obsessed him for so many years.

The joint French-Tuscan expedition, led by Jean-François Champollion with Ippolito Rosellini as second-in-command, consisted of twelve other men, many of whom were artists and draughtsmen, indicating the great importance attached to the recording of monuments. From France, the expedition members were the Egyptologist and numismatist Charles Lenormant, the architect

Antoine Bibent, the traveller and artist Alexandre St. Romain Duchesne, the archaeologist and draughtsman Nestor L'Hôte, and the artists Édouard François Bertin and Pierre François Lehoux. The Tuscan contingent, headed by Rosellini, consisted of his uncle Gaetano Rosellini, who was an engineer and architect, the artists Salvatore Cherubini and Giuseppe Angelelli, the botanist Giuseppe Raddi with his assistant Galastri, and the explorer and physician Alessandro Ricci, who had previously travelled in Egypt.

The sight of Alexandria amazed everyone, with its forest of masts belonging to so many vessels, especially the French and English ships blockading the port: 'This mixture of vessels of every nation, friends and enemies all together, is a most remarkable sight and suffices to characterize these times.' The first task was to contact the French and Tuscan consuls. The French Consul Drovetti was astonished to see Champollion who, being informed of Drovetti's letter warning the expedition not to set sail, remained philosophical: 'My good star shone on this such important occasion, because the letter didn't reach me in time.' Although still pessimistic about the political situation, Drovetti agreed to begin the process of obtaining the necessary permits from the ruler of Egypt, Mehemet Ali.

While waiting for the permits to travel through Egypt and Nubia and undertake excavations, the expedition members began to explore Alexandria, and like the savants of Napoleon's expedition they were soon drawn to the obelisks called 'Cleopatra's Needles.' Thirty years before, the savants could do little more than admire the obelisks as monuments, but now Champollion could read the hieroglyphs, noting that although later inscriptions of Ramesses II were present, the earliest ones showed that the obelisks had originally been erected by Tuthmosis III in front of the temple of the sun at Heliopolis, over 100 miles away. Copies were made of the inscriptions on these obelisks, even though they had already appeared in a volume of the *Description de l'Égypte*. On comparing the published versions with the actual inscriptions many errors became obvious, and in a letter to his brother Champollion wrote: 'Bibent is finishing off the three sides of the obelisk which were known to the Commission...which nobly mangled the hieroglyphic inscriptions.'

Napoleon's savants may not have made much impression on the hieroglyphic inscriptions of Egypt, but they had not been forgotten by the Egyptians. Some of the older Egyptians could still speak French, and Champollion was delighted that Joseph Fourier in particular was well remembered as fair and generous in his role as an administrator. On one occasion Champollion was met by a blind Arab who addressed him in French with the words: 'Good-day citizen, give me something, I have not yet eaten.' Amazed by this Republican style of speech that must have originated from Napoleon's expedition, he gave some French coins to the old man, who on identifying them by touch said, 'That no longer passes, my friend!' — and so Champollion changed them for Egyptian coins. Later he recorded in his journal: 'One finds at each moment in Alexandria old reminders of our campaign in Egypt.' The draughtsman L'Hôte noticed that many Egyptians fled from them, believing them to be tax collectors, but they were welcomed when it was realized that they were French because 'the recollection of the memorable expedition of Bonaparte has not yet been entirely wiped out with the poor Arabs who each, at that time – these were their own words – had their own donkey and their own cow, and didn't pay taxes twice.'

With few other monuments or artifacts with hieroglyphic inscriptions to study in Alexandria, at least Champollion found that observations of everyday life in this Egyptian town unravelled some of the remaining problems of understanding hieroglyphs. Not being able to avoid seeing the packs of stray dogs, he compared them with hieroglyphs: 'The DOG lives in Egypt in a state of complete freedom ... They greatly resemble the *jackal*, except for the coat which is reddish-yellow. I am no longer astonished that in the hieroglyphic inscriptions, it is so difficult to distinguish the *dog* from the *jackal*!' In fact, four hieroglyphs can easily be confused: 𓃡 (greyhound), 𓃥 (recumbent dog), 𓃩 (jackal) and 𓃨 (possibly a wolf on a standard).

The weeks spent waiting for permits did allow the expedition members to acclimatize themselves to the country and its customs; they gradually became conditioned to the heat, riding donkeys and resting during the hottest part of the day. In order not to be too conspicuous, they had abandoned European costume for Egyptian clothes: an imposing turban on top of a shaved head, an embroidered jacket over a striped silk waistcoat, baggy trousers with a wide waistband and scimitar, and soft shoes or slippers – with his swarthy complexion and faultless Arabic, Champollion could pass for a native of the country. They also employed the time in collecting the equipment and supplies that would be needed for the journey up the Nile and in preparing the two boats that would carry them.

As the days passed without the permits arriving it became obvious that a deliberate attempt was being made to delay the expedition, and Champollion found out that the antiquities merchants were being obstructive, because they had 'all quivered at the news of my arrival in Egypt with the intention to excavate.' As Drovetti had close connections with the antiquities trade, Champollion suspected his involvement and also began to wonder if this was the real reason for the letter warning the expedition not to come to Egypt. He forced the issue by pointing out to Drovetti that the expedition had the full backing of King Charles X and that if the permits were not forthcoming, he would be obliged to explain to the King how his scientific expedition had been blocked because of the personal interests of a handful of antiquities dealers. After this ultimatum, the permits arrived just a few days later.

The expedition's two boats were named *Isis* and *Hathor*, after the two most important goddesses of ancient Egypt, and on 14 September the expedition members, along with domestic help, crew and two policemen provided by Mehemet Ali, set off along the Mahmoudieh Canal, which had been cut across the desert only nine years earlier to connect Alexandria to the Rosetta branch of the Nile. Champollion's plan of campaign was to journey south up the River Nile as far as the second cataract, stopping only to identify sites and to copy the most important inscriptions and sculptured reliefs – having reached the second cataract and assessed the relative importance of the places they had visited, the return journey would be spent making a detailed study of the best sites in the time available. The expedition was also to look out for suitable antiquities with which to enhance the Egyptian collections of France and Tuscany. After nearly a day of travelling, they left the Mahmoudieh Canal and were at long last on the Nile – having tasted it, Champollion declared the river a 'champagne among waters.'

Sailing south on the Rosetta branch of the Nile they proceeded past the village of Desouk, where Champollion was reminded of Henry Salt and his wonderful collection of antiquities that he had succeeded in obtaining for the Louvre, noting sadly in his journal: 'I learned that it was in a country house within the neighbourhood of this village, on the eastern bank of the Nile, that M. Salt, consul general of England, died some months previously. I still regret no longer finding in Egypt this educated man and great lover of hieroglyphic studies.' The next day the expedition halted briefly to examine its first ancient site along the Nile – the town of Sais, centre of the cult of the goddess Neith. A desolate area, Sais was strewn with masses of broken pottery and was largely flooded by the Nile, with an all-pervading stench from the modern cemetery that gave rise to a heated discussion among the members of the expedition about attempts to control the plague in Egypt.

Continuing to stop at any place that gave a hint of ancient remains, and noting in passing the sites of Napoleon's battles with the Mamelukes, it took some five days to come within reach of Cairo and the first glimpse of the distant pyramids, a view vividly described in a letter to Jacques-Joseph: 'In waking up on the morning of the 19th we finally saw the pyramids whose massive shape we could already appreciate although we were at a distance of eight leagues. At 1.45 p.m. we arrived at the summit of the Delta (*Bathn-el-Bakarah*, The Belly-of-the-Cow), at the very place where the river splits into two great branches, that of Rosetta and that of Damietta.

The view is magnificent, and the breadth of the Nile astonishing. To the west, the pyramids rise up in the middle of palm trees; a multitude of small boats and vessels criss-cross in every direction. To the east...the bottom of the picture is occupied by the *Muqattam Hills*, which encircle the citadel of Cairo, and whose base is hidden by the forest of minarets of this great capital.'

Having passed the site of the Battle of the Pyramids and offered a greeting to the shades of Napoleon and his soldiers, the expedition entered Cairo the next day. The city was in a joyful uproar for a celebration of the anniversary of the Prophet Mohammed, and the expedition was given an unexpectedly warm welcome – Champollion was astonished to find that some already knew him as 'the man who could read the writing of the old stones.' The next day they began to explore Cairo, and in contrast to Napoleon's troops, Champollion found it an enchanting place:

People have spoken very badly of Cairo: for me I find it rather fine, and these streets of eight to ten feet in width which are so criticized appear to me perfectly well calculated for avoiding the extremely great heat...Cairo is an absolutely monumental city...A multitude of mosques, some more elegant than others, covered with the most tasteful arabesques and decorated with wonderful minarets of richness and charm, give to this capital an imposing and very varied appearance...Cairo is still a city of One Thousand and One Nights.

The state of his health had been improving steadily since arriving in Egypt, as he explained to Jacques-Joseph: 'My health is still excellent and better than in Europe because I have written these seven pages to you in one go, which I might have been incapable of doing at Paris without spasms in my head. It is true that I am quite a new man.' All too aware of the long journey and exertion that lay ahead, he gave the expedition members their freedom in the city – a short holiday before the hard work began – but he always felt the pressure of passing time, and after ten days in Cairo the boats were loaded with fresh provisions and the expedition headed south.

The first stop was just up the river from Cairo at Toura, a huge area of limestone quarries about six miles in circumference that had been used as a source of stone since prehistoric times. Here the expedition first used a system that would become instinctive as they moved from site to site up the Nile Valley. Each member of the expedition was allotted an area to search and record, and Champollion would be summoned to examine anything thought to be of particular interest, concentrating especially on inscriptions: 'I went to each place in order to appreciate the importance of the discovery. If the inscription appeared interesting, I drew it, or had it drawn if it was formed of very distinct lines.' There were many demotic and some hieroglyphic inscriptions at Toura, and the copying of the inscriptions in the intense heat was a harsh task done as rapidly as possible.

Moving to the west bank of the Nile, the expedition went to the area of the ruins of the city of Memphis, for a time the capital of ancient Egypt, which was now partially flooded owing to the Nile inundation. This annual event was a result of the heavy summer rains in Ethiopia, which reached Aswan in late June and Cairo at the end of September, when the waters would begin to recede. The effect of these floods was to deposit fertilizing black silt down the length of the Nile Valley. In ancient Egypt the height of the inundation governed whether there would be surplus food or a famine in the following year, so its level was monitored up and down the Nile by 'Nilometers,' and the god Hapy was worshipped as a personification of the flood. The annual inundation, the pulse of ancient Egypt, has not happened for thirty years because the floodwater is now stored in the vast Lake Nasser behind the Aswan Dam in the south of the country.

Being in Memphis at the beginning of October, the expedition was hampered by the full effects of the flooding. The area that was dry mainly consisted of granite blocks strewn across the sand, but it was possible to record the remains of the colossal statue of Ramesses II, one of the few remnants of the desolate desert site that is visible today. At nearby Mît-Rahineh, a village that encroached upon the site of the ancient city, the expedition carried out excavations in which a burial ground and temples dedicated to Hathor were discovered. The threat of further flooding cut short the excavations, and they took their tents and equipment on a train of camels to Saqqara two miles away, as it was too far from the Nile for them to return to the boats to sleep. While setting up camp, they met a group of Bedouin, and established very good relations with them (in stark contrast to Napoleon's expedition), hiring several as labourers and night watchmen. With his open-minded sense of justice, Champollion commented, 'these are brave and excellent people, when one treats them as men.'

The expedition was disappointed by its first sight of Saqqara, which had been the main necropolis of the city of Memphis for the kings and the elite. Expecting sumptuous buildings, numerous pyramids, temples, tombs and avenues of sphinxes, they found instead the debris left by looters who had systematically turned the site upside down over the previous twenty years, and in many places the debris was covered by drifts of sand. Champollion lamented the situation in a letter to his brother: 'I visited here, at Sakkara, the plain of the mummies, the ancient cemetery of Memphis, strewn with violated tombs and pyramids. This locality, thanks to the barbarous greed of the merchants of antiquities, is completely empty for study.' It was here too that the architect Bibent left the expedition, because his health had deteriorated from the start, and he was no longer full of the energy and enthusiasm Champollion had noted and applauded when he first met him – in fact, Bibent barely survived a year after leaving Egypt.

One topic that did not find its way into Champollion's letters (many of which Jacques-Joseph was to publish in France in an edited form) was the significance of the names of pharaohs he found at Saqqara.

The above cartouches (Djedkare and Isesi) gave him the names of a pharaoh of the fifth dynasty who reigned for about forty years from around 2414 B.C. He was not certain of the exact dates of this pharaoh, but it was enough to confirm the existence of a pharaoh of the fifth dynasty. Scholars knew about early dynasties of kings of Egypt from the Greek writings of Manetho, a late third-century B.C. historian and Egyptian high priest. In his history of Egypt (the *Aegyptiaca*), Manetho gave a chronology of the country, beginning with the gods, demigods and spirits of the dead, followed by the Flood and then thirty dynasties of kings, a system of listing the pharaohs which was adopted by Champollion and is still used today. Corresponding fairly closely to other more ancient lists of kings, such as the Turin Royal Canon discovered by Champollion, the divisions between Manetho's dynasties seem to have been decided by the places from which particular groups of kings ruled, such as Memphis and Thebes. As Biblical chronology denied the existence of the first fifteen dynasties of Manetho, because they were far too early (and therefore could not have occurred), the discovery of the name of a pharaoh of the fifth dynasty was in direct and terrible conflict with that chronology, throwing immediate doubt on the date of the creation of the world. Champollion was to find more and more evidence that clashed with theological dogma, but he was usually careful to confine the facts within his personal records and journals, not even daring to discuss the findings with other members of the expedition.

Against expectations, one of the tombs at Saqqara provided good evidence for the ancient Egyptian calendar, and Champollion noted in his journal that the Egyptians divided the night into twelve hours and the day into twelve hours, correctly identifying 🕊️○✶ as 'hour' (more usually written as 🕊️○✶). In fact, the Egyptians appear to have been the first people to use this division, which gives twenty-four hours in the day. The year itself was divided into three seasons: 𓈌 (*akhet* – the season of the Nile flood), 𓉐𓇳 (*peret* – the equivalent of spring, when the crops began to emerge), and 𓈖𓇳 (*shemu* – harvest time). Each season was divided into four months, and each month was divided into three weeks of ten days, so that the year consisted of twelve months, each thirty days long, giving 360 days in the year. To bring the number up to 365, five days were added that were regarded as the birthdays of the deities Osiris, Isis, Horus, Seth and Nephthys. Dates were given according to the regnal year of the current pharaoh. For example, 𓇳𓏏𓏏𓏤 𓇳𓏺 means 'year two, second month of inundation, day one under the Majesty of King Khakaure' – Khakaure was the throne name of the twelfth dynasty Pharaoh Senusret III, who probably reigned from about 1878 B.C. to 1841 B.C.

From the frustrations of Saqqara, the expedition caravan moved on to the pyramids and the Sphinx at Giza, ten miles to the north-west, another major necropolis, now on the very edge of Cairo's suburbs. Like many modern visitors to the site Champollion suffered the illusion that the three large pyramids lost their grandeur as he approached, recording the strange sensation in his journal:

Everyone will be as surprised as I that the impression of this prodigious monument decreases as you move closer to it. I was myself in some way humbled on seeing, without the least astonishment, at a distance of fifty paces, this construction whose calculations alone can make one appreciate its immensity. It seems to sink as one approaches, and the stones which form it only appear as rubble of a very small size. It is absolutely necessary to *touch* this monument with one's hands in order to at last become aware of the enormous size of the materials and of the massive shape that the eye is meas-uring at that moment. At a distance of ten paces the hallucination [that it seems small] resumes its power, and the

great Pyramid appears nothing more than a common building. One truly regrets having been close to it.

The Sphinx, on the other hand, remained impressive through
being enigmatic. The draughtsman L'Hôte noted that: 'This monument, that one knows represents a symbolical being with the body
of a lion and a human head, is buried right up to the top of its shoul-
ders in the sand, through which one can trace the form of the back
and hindquarters of the animal.' After three days of exploring the
area around the Giza pyramids and recording scenes and inscriptions inside the tombs, the expedition returned to the boats for the
long haul upriver to Beni Hassan. It took twelve days to cover the
125-mile distance, and the expedition arrived late in the evening of
23 October. Previous travellers had reported that the site with its
tombs cut into the limestone cliffs was insignificant, and Champollion expected to complete all necessary work in one or two days. The
initial examination of the interiors of the tombs revealed little in the
way of decorations or inscriptions except for a few tantalizing
marks under a thick coat of dust, but Champollion thought to clean
off some of the dust with a wet sponge. To everyone's amazement
vividly colourful paintings were revealed beneath, depicting all
manner of subjects such as agriculture, arts and crafts, military
scenes, games and musicians, singers and dancers – wonderful pictures that would take far more than a day or two to record.

When the expedition finally left Beni Hassan in early November
it was lagging behind the schedule that Champollion had planned,
as he noted to Jacques-Joseph:

All this is the fault of the admirable Jomard, who, describing
the rock-cut tombs of this locality, gave such a meagre idea of
them by his little inexact drawings and his rather more doubtful language, that I was counting on dealing with these caves
in *one* day; but they have devoured fifteen... This life of tombs
has had as a result a portfolio of drawings perfectly made and

of a complete exactness, which already exceed more than 300. I dare say that with these riches alone, my voyage to Egypt would be already better fulfilled and more productive than all the papers of the *Commission*.

Not all the paintings and inscriptions in the tombs were Egyptian, since various travellers had recorded their presence with graffiti. In one tomb was an inscription made by one of Napoleon's soldiers which simply read '1800, 3rd regiment of dragoons.' Champollion reverently went over the inscription in ink to make it more readable and added his own mark below: 'J. F. C. Rst. 1828.'

As well as providing such a wealth of material, the surprising richness of the tomb paintings at Beni Hassan had other less favourable consequences. Wanting to press on towards Thebes, Champollion had driven both himself and the other expedition members to finish recording the tomb paintings as quickly as possible so as to minimize the disruption to his plan of campaign. The result was a deterioration of his own health as well as disaffection among some of the others – it was a grumbling group of exhausted men that eventually left Beni Hassan with over 200 miles to travel before they saw Thebes, and at least as much again before they reached the second cataract.

Although time was pressing, the expedition still visited the major sites, but kept to an absolute minimum the time spent at each, a job unfortunately made easier by the discovery that some had been destroyed. At the Roman town of Antinopolis, founded in the second century by the Emperor Hadrian in memory of his lover Antinous who had drowned in the Nile, the columns, baths, triumphal arches and porticoes of the hippodrome and theatre that had been recorded by Napoleon's savants had all disappeared, much to the disgust of Champollion: 'None of the monuments described by the *Commission of Egypt* has escaped the fury of the barbaric inhabitants, who, with the permission of their government, have destroyed everything, right down to the foundations.' With their hurried progress, they managed to reach Dendera in just over a week, arriving after dark on 16 November.

From the Nile it was not possible to see Dendera, the site which had had such a remarkable effect on Napoleon's troops and on the artist Vivant Denon, and from where the controversial zodiac had been taken to Paris, but in the brilliance of a moonlit night the thought of the site exerted a powerful draw on the expedition members. Champollion recorded their compulsive excitement in a letter to Jacques-Joseph:

> The moonlight was magnificent, and we were only at a distance of one hour from the temples: could we resist the temptation? I ask this of the coldest of mortals! To eat and to leave immediately was the work of a moment: alone and without guides, but armed to the teeth, we set off across the fields, presuming that the temples were in direct line with our boat. We walked like this, singing the most recent opera marches, for an hour and a half, without finding anything. Finally a man was discovered; we called to him, and he bolted, taking us for Bedouin, for, dressed in the eastern manner and covered with a great white hooded cape, we resembled to the Egyptian man a tribe of Bedouins, whilst a European might have taken us without hesitation for a guerilla force of Carthusian monks armed with guns, sabres and pistols. The runaway was brought to me and...I ordered him to lead us to the temples. This poor devil, barely reassured at first, took us along a good route and finished by walking with good grace: thin, dry, black, covered with old rags, this was a *walking mummy* but he guided us rather well and we treated him in the same way. The temples finally appeared to us. I will not try to describe the impression which the great propylon and especially the portico of the great temple made on us. One can measure it easily, but to give an idea of it is impossible. It is grace and majesty brought together in the highest degree.

They returned to the boats to sleep at three o'clock in the morning, only to rise again four hours later to explore the site in daylight. Now able to see all the wonderful sculptured reliefs clearly and with his understanding of hieroglyphs, Champollion noted that it was a Ptolemaic temple, dedicated not to Isis as had previously been thought, but to Hathor – it is at times difficult to distinguish between some deities, particularly Isis and Hathor, unless the accompanying hieroglyphs can be read. Champollion regarded the temple as a masterpiece of architecture covered in decadent sculpture, but what really amazed him were the empty cartouches –
⬭ – devoid of names or titles of rulers:

> in the whole interior of the naos, as well as in the rooms and the buildings constructed on the terrace of the temple, there doesn't exist a single carved cartouche: all are empty and nothing has been erased. The most pleasing thing of this business, *risum teneatis, amici!* [you may restrain your laughter, friends!], is that the piece of the famous circular zodiac which bore the cartouche is still in place, and that this same cartouche is *empty*, like all those of the interior of the temple, and never received a single blow from the chisel. It was the members of the *Commission* who added to their drawing the word *autocrator*, believing that they must have forgotten to draw an inscription, which didn't exist.

For years arguments had raged about the date of the Dendera zodiac, with Young identifying an accompanying cartouche, published by the Commission of Egypt, as that of Arsinoë. Champollion had put an end to the debates by correctly identifying the published cartouche as 'Autocrator,' a title used in Roman times, so apparently dating the zodiac. Now he found that the Commission had used an inscription from elsewhere to fill up the empty cartouche in their publication. He appreciated the irony that his dating of the Dendera zodiac, which had earned him the reputation of defender of the Biblical chronology, the favour of the Pope and indirectly the award of Knight of the Legion of Honour, had been based on a forged drawing.

Despite the wonders of Dendera, Thebes was now so close that Champollion was ever more anxious to move on, and they arrived after three days sailing against contrary winds: '*Thebes!* This name was already very great in my thoughts: it has become colossal since I went all over the ruins of the old capital, the oldest of all the towns of the world. For four whole days I have raced from marvel to marvel.' Of the four days devoted to the huge number of ruins on both the east and west banks of the Nile, Champollion used the first to explore the temple of Ramesses II as well as the colossal statues that were the sole remains of the temple of Amenhotep III. The second day was spent at the temples of Medinet Habu and the third day amongst the tombs. The spectacular sights of the last day spent at Luxor and Karnak were described in a long, passionate letter to Jacques-Joseph:

On the fourth day...I went from the left bank of the Nile in order to visit the eastern part of Thebes. First of all I saw *Luxor*, an immense palace in front of which were two obelisks nearly eighty feet in height, of a single block of rose granite, of an exquisite workmanship, accompanied by four colossal statues of the same material and thirty feet in height approximately, for they are buried right up to their chests...I finally went to the palace or rather to the town of monuments, to *Karnak*. There all the pharaonic magnificence appeared to me; everything that men have imagined and carried out on a grander scale. Everything which I had seen at Thebes, everything which I had admired with enthusiasm on the left bank, appeared to me wretched in comparison with the gigantic ideas with which I was surrounded...we in Europe are only Lilliputians, and no ancient or modern people has conceived the art of architecture on such a sublime, large, and grandiose scale as the old Egyptians.

Moving just ten miles south-west from Thebes, the expedition spent a day recording the inscriptions at the Ptolemaic temple at Hermonthis, a temple that was later dismantled and its masonry burnt in lime kilns – the expedition's records are almost the only evidence of the existence of this site. Farther upriver they had cause to lament the delays that had occurred in their journey southwards, for only twelve days before their arrival a temple near Esna had been demolished and the stone used to reinforce the nearby Nile barrage that was in danger of being carried away in the annual flood. Pressing on farther south they were relieved to find that the temple of Edfu was still intact, although masked by heaps of rubble and Arab huts clustered around and on top of it. Having recorded those inscriptions that were accessible, Aswan was eventually reached on 4 December.

The first cataract at Aswan effectively marked the border between the countries of Egypt and Nubia, forming a physical barrier that forced the expedition to leave their two boats and transfer their supplies and equipment to several smaller boats south of the cataract, in order to continue the journey. Aswan itself was of interest, being the ancient town of Syene, and Champollion especially wanted to see the two temples on Elephantine Island opposite Aswan, but was bitterly disappointed to find that like many other sites they had been recently dismantled and the stone reused – on this occasion to build a military barracks and a new palace for Mehemet Ali.

At Aswan Champollion's health once again troubled him, with such a severe attack of gout that he could only walk with two men to support him, and it was in this fashion that he made a painful visit to the temple of Isis on the nearby island of Philae. Unknown to him it was on a Roman gateway there, fourteen centuries earlier, that the last dated hieroglyphic inscription was carved into stone on 24 August A.D. 394. He did manage to record a commemorative plaque set up by French troops on 3 March 1799 to mark the southernmost point reached by Napoleon's Egypt expedition, an achievement that Champollion and his companions were about to surpass by travelling south through Nubia to the second cataract. At this stage, Champollion's health was so fragile that after the effort of the visit to the Philae temple he needed several days of rest to recover, but by 16 December 1828 everything was loaded aboard a squadron of seven boats above the cataract, and the expedition left Egypt and set out into Nubia.

Keeping fairly well to the plan of stopping as little as possible on the way south, they reached the two rock-cut temples at Abu Simbel after ten days, where Champollion was anxious to see at first hand the inscriptions whose drawings (done by the architect Huyot) had given him such vital clues to the principles of the hiero-glyphic script over six years earlier. The expedition spent two days exploring the two immense temples, concentrating on the biggest temple once huge quantities of sand had been removed, allowing a perilous access as Champollion described in a letter to Jacques-Joseph:

The great temple of Abu Simbel is alone worth the journey to Nubia: it is a marvel which would be a rather beautiful thing even at Thebes. The labour that this excavation cost terrifies the imagination. The façade is decorated with four seated colossal statues, no less than sixty-one feet in height. All four, of superb workmanship, represent Ramesses the Great: their faces are *portraits* and perfectly resemble the figures of this king which are at Memphis, at Thebes and everywhere else. It is a work worthy of all admiration. Such is the entrance; the interior is completely worthy of it, but it is a rough job to visit it. On our arrival the sand, and the Nubians who take care to shift it, had closed the entrance. We had it cleared so as to ensure as well as possible a small passage that they had opened, and we took all precautions possible against the infernal flow of this sand which, in Egypt as in Nubia, threat-ens to engulf everything. I almost completely undressed, only keeping my Arab shirt and underpants of cotton, and went forward flat on my stomach to the little opening of a door, which, if cleared, would be at least twenty-five feet in height. I thought I was going forward into the mouth of an oven, and sliding entirely into the temple, I found myself in an atmosphere heated to fifty-two degrees: we rushed through this astonishing excavation, Rosellini, Ricci and I and one of our Arabs, each holding a candle...After two and a half hours of admiration and having seen all the reliefs, the need to breathe a little pure air made itself felt, and it was necessary to get back to the entrance of the furnace.

Outside the temple Champollion resumed his layers of clothing (used as insulation against the heat at that period), including two flannel vests, a greatcoat and burnous, and sheltered from the keen wind by sitting next to one of the colossal statues of Ramesses II in order to regain his strength. He was still sweating profusely from the effort of exploring the temple when he returned to the boat, but his mind was reeling with the visions of the beautiful reliefs and he was determined to have copies – he 'would do everything to have them.'

Curbing his impatience, Champollion led the expedition from Abu Simbel on the short journey to Wadi Halfa, just below the second cataract and now within modern Sudan. If they were to go farther south from here, they knew they would have to leave the boats and much of their supplies and walk across the desert, but at that time a famine was raging in Nubia, and it was unlikely that the expedition could travel any significant distance without running dangerously short of food, so the decision was taken to keep to the plan and return back down the Nile, studying the important sites more thoroughly. The expedition spent the New Year at Wadi Halfa, and for Champollion at least it was a time of reflection on what they had achieved and what lay ahead. In the four months that he had been in Egypt, his system of decipherment had been tested against all manner of inscriptions – not only had it worked, but it was being constantly refined and enlarged. Although some of his rivals still denied it, there was no longer any room for doubt that the hiero-glyphs had been deciphered and that the ancient Egyptian texts could once more be read.

Champollion wrote several letters at Wadi Halfa that, in their dif-ferent ways, reflected his thoughts, hopes and fears at this time. In a letter to Dacier, he summarized what the expedition had achieved so far and how his system of decipherment was working:

I am now proud that, having followed the course of the Nile from its mouth right up to the second cataract, I have the right to announce to you that there is nothing to change in our *Letter on the alphabet of hieroglyphs*. Our alphabet is good: it has been applied with equal success first of all to the Egyp-tian monuments of the time of the Romans and the Greeks, and afterwards, which becomes of much greater interest, to the inscriptions of all the temples, palaces and tombs of the pharaonic times. Everything therefore justifies the encour-agement that you wanted to give to my hieroglyphic work at a time when nobody was disposed to lend it favour...Philae has been almost exhausted during the ten days that we spent there while going up the Nile, and the temples of Ombos, Edfu and Esna, so flaunted by the *Commission of Egypt* to the detri-ment of those of Thebes, that these Gentlemen didn't *appreciate*, will stop me for a little time...My portfolios are already very rich: I look forward in advance to putting before your eyes all of old Egypt, religion, history, arts and crafts, customs and manners. A great part of my drawings are col-oured, and I am not afraid of announcing that they look nothing like those of our friend Jomard, because they repro-duce the true style of the originals with a scrupulous fidelity.

In writing to his brother, Champollion was less reserved about what lay ahead: 'My work *really begins today*, although I already have in the portfolio more than 600 drawings, but there remains so much to do that I am almost frightened by it.' In the same letter he told Jacques-Joseph that he was relying on him to arrange a French naval vessel to pick them up at Alexandria at the beginning of October, and in a letter to his old friend Thévenet he commented wryly: 'My health has held up, and I hope that that will continue – I am sober as much by necessity as by virtue, and, one helping the other, I will avoid the illnesses of the country.'

The Egyptologist Charles Lenormant had agreed with Champollion to accompany him only as far as the second cataract, and although a loyal member of the expedition, Lenormant now held to that agreement, heading back north in one of the boats. With the early loss of Bibent, the expedition was now down to twelve men. As they turned back down the Nile from Wadi Halfa on 1 January 1829, the boats stopped the next day below the Cave of Machakit, situated high up on a cliff face. Only Champollion, L'Hôte and Ricci risked the dangerous climb to the cave, which had been made into a chapel during the reign of the Pharaoh Horemheb, around 1300 B.C., and while Champollion copied the inscriptions, the other two drew the sculptured reliefs. The descent from the cave was made more dangerous by the onset of a severe storm, but the three men reached the bottom of the cliff safely and the boats immediately cast off. They had not travelled for more than half an hour when the violence of the wind forced the boats to the bank, where they were obliged to remain until the storm abated in the early hours of the following day.

The day after the storm found Champollion with a bad attack of gout in his right knee that forced him to take to his bed, and on arriving at Abu Simbel he was irritated at not being able to make copies of the inscriptions himself. While the other members of the expedition carried on the necessary drawing and copying both inside and outside the main temple, he passed the time compiling notes for his hieroglyphic dictionary and listening sympathetically to the lamentations of the naturalist Raddi, who had just discovered that his extensive collection of rock samples had been dumped in the Nile by one of the sailors, who feared that his boat would sink. On 6 January, still not properly recovered, Champollion insisted on working in the temple and had to be almost carried there. Once inside, the sight of all the statues, reliefs and inscriptions seemed to lend him strength, enabling him to work for some two hours. The humid heat within the temple helped relieve his gout, and over the next few days he worked there for about three hours at a time while his health gradually improved.

All the members of the expedition working inside the temple found it a gruelling experience, but the information they were recording was too important to ignore. Champollion described the problems to Jacques-Joseph, while applauding the brilliant work of his colleagues:

> Everything is colossal here including the work that we have undertaken, whose result will have some claim to public attention. All those who are familiar with the locality know what difficulties one has to conquer in order to draw a single hieroglyph in the great temple...and when you hear that the heat that one experiences in this temple, today *underground* (because the sand has almost covered its façade), is comparable to that of a strongly heated Turkish bath, when you hear that it is necessary to enter it almost naked, that the body perpetually perspires with an abundant sweat which runs down the eyes, drips onto the paper already soaked by the humidity of this heated atmosphere like in an oven, one will admire without doubt the courage of our young people who brave this furnace for three or four hours each day, who only leave through exhaustion and only quit the work when their legs refuse to carry them.

While the artists and draughtsmen drew everything and anything, it was left to Champollion and Rosellini to maintain the accuracy of the copies of the hieroglyphic texts:

Rosellini and I, we have reserved for ourselves the task of the hieroglyphic inscriptions, often very extensive, which accompany each figure or each group in the historical reliefs. We copy them on the spot, or after prints on paper when they are placed at a great height; I check them several times against the original, I do good copies of them, and immediately give them to the draughtsmen who, beforehand, have marked out and traced the columns destined to receive them.

It took thirteen days of exhausting labour for the expedition to make all the necessary detailed records of the two temples at Abu Simbel, under the most difficult conditions that they were to encounter, feeling at the end of that time that they were really on the homeward journey – Champollion wrote in his journal: 'I have not been able to hold back a feeling of sadness when leaving forever in this way…this beautiful monument, the first temple which I am leaving never to see again.'

It was less than one day's sailing from Abu Simbel to Qasr Ibrim, a strong citadel, set high on a rocky promontory, that had been destroyed some years before by Mehemet Ali's troops to deny its use to the Mamelukes. The expedition stopped briefly to examine the caves at the foot of the cliffs, which could only be reached with the aid of ladders. These caves proved to be rock-cut shelters and chapels dating back at least to the eighteenth dynasty, around 1500 B.C. With the building of the Aswan Dam in the 1960s a vast area including Qasr Ibrim was flooded by Lake Nasser – now the water of the lake laps against the ruined walls of what was once a formidable citadel, and the caves are many metres below the surface. Much of Nubia was destroyed by the building of the dam, and although some of the sites visited by Champollion's expedition were rescued by a massive UNESCO project involving over forty countries, many other sites were not moved and are now inaccessible beneath the deep waters of Lake Nasser. The most spectacular rescue was that of the two rock-cut temples at Abu Simbel, which were moved to a new location within an artificial rocky outcrop constructed over a concrete dome.

Having arrived at Qasr Ibrim in the early morning, the expedition pressed on and by the evening of the same day reached Derr, the capital of Nubia, described by a contemporary guide for travellers as a 'long, straggling village of mud cottages…and a mosque – the only one seen after leaving Philae' and by Champollion as 'a large village of two hundred houses, but more agreeable and cleaner than many of the towns of Egypt, because the streets are spacious and especially because the houses are surrounded by little plantations of palm trees.' After supper by moonlight, Champollion spoke with one of the inhabitants, asking him 'if he knew the name of the *sultan* who had had the temple of *Derri* [Derr] constructed; he immediately answered me that he was too young to know that, but that all the old men of the country had appeared to him to agree that this *Birbé* [temple] had been constructed around three thousand years before Islam, but that all the old men were uncertain on one point – knowing if it was the *French*, the *English* or the *Russians* who had at that time carried out this work.' Enchanted by this explanation, Champollion commented: 'Here is how one writes history in Nubia.'

At sunrise the following day a visit was made to the rock-cut temple of Derr, which served to solve one of the minor problems that had been nagging him. A lion (the symbolic animal for which he still retained an affection) was shown in some carved reliefs accompanying the Pharaoh Ramesses II into battle, but Champollion had been unable to determine whether this was merely a symbolic association indicating that Ramesses had the courage and strength of a lion in battle, or whether it was a real lion tamed and trained for war. At Derr the Pharaoh's lion was actually portrayed throwing itself on his enemies with the caption: 𓃭𓍯𓈖𓂝𓏏𓊃𓊪 𓅭𓃭𓈖𓏤𓏤𓏤 ('The lion, servant of His Majesty, tearing his enemies to pieces'), to which Champollion commented: 'That seems to me to demonstrate that the lion really existed and followed Ramesses into the battles.'

Continuing northwards, a series of temples came under scrutiny as they moved downriver, until they arrived at Philae in the evening of 1 February. Here they spent six days while their equipment was hauled around the first cataract and loaded back onto their original boats, the *Isis* and *Hathor*, at Aswan, from where the expedition worked its way back down the Nile recording various sites, finally returning to Thebes on 8 March. The first two weeks at Thebes involved a detailed study of the Luxor temple on the east bank of the Nile, in front of which the boats were moored and used for sleeping, because no other suitable accommodation was available. The expedition next planned to explore the burial ground of the pharaohs on the west bank – the ancient Egyptians had several names for this necropolis, including 'Beautiful Place,' 'Great Field' and 'Beautiful Ladder of the West,' but its formal name was:

meaning 'The Great and Noble Tomb of the Millions of Years of the Pharaoh, Life, Strength, Health, in Western Thebes.' Champollion learned that the Arabic name for the burial ground was Biban-el-Molouk meaning 'Gates of the Kings' – an unconscious echo of the 'Thebes of a Hundred Gates' as the Egyptian Thebes is described in Homer's *Iliad* in order to distinguish it from Thebes in Greece. Adapting the name slightly, Champollion called the place The Valley of the Kings, the name by which it is most commonly known today.

The expedition took up residence in the tomb of the Pharaoh Ramesses IV, which had long since been completely robbed and occasionally used as shelter by previous travellers. In a letter to his brother a few days later, Champollion described their unusual accommodation:

> Our caravan, composed of donkeys and savants, therefore set itself up here on the same day, and we are occupying the best and most magnificent lodging that it is possible to find in

Egypt. It is King Ramesses (the 4th of the 19th dynasty) who gives us hospitality, for we are all living in his magnificent tomb, the second that one meets on the right when entering the Valley of Biban-el-Molouk. This rock-cut tomb, admirably preserved, receives enough air and enough light that we lodge there marvellously. We occupy the first three rooms, which form a length of sixty-five paces; the walls, from fifteen to twenty feet in height, and the ceilings are all covered in painted sculptures, whose colours preserve almost all their brilliance. It is a true habitation of a prince...Such is our establishment in the *Valley of the Kings*, a true resting place of the dead, because you find here neither a blade of grass, nor living beings, with the exception of the jackals and hyenas, who–the night before last–devoured at a hundred paces from our *palace* the donkey which had carried my servant Mohammed.

With his letter Champollion enclosed a plan of the tomb showing how its long entrance tunnel was divided into rooms and where everyone slept. The plan confirms his identification of the tomb as that of Ramesses IV – for some reason, perhaps because it is easy to confuse the Roman numerals IV and VI, the expedition was later incorrectly recorded as occupying the tomb of Ramesses VI. Unwittingly, they had established themselves in the tomb whose ancient Egyptian plan Champollion had discovered amongst the papyri of the Drovetti collection at Turin. Because the ancient plan was incomplete and the only modern survey of the tomb available to him was an inaccurate one drawn by the Commission of Egypt, he failed to realize this remarkable coincidence. It was only many years later, when an accurate survey was being made by Howard Carter (before he became famous for discovering the tomb of Tutankhamun exactly 100 years after Champollion's hieroglyphic breakthrough in Paris), that the tomb of Ramesses IV was proved to be the one drawn on the ancient plan.

In the Valley of the Kings they meticulously recorded the paintings and inscriptions of the sixteen tombs that were partially or wholly accessible for study and had the best paintings. Because of damage in more recent decades from water seepage, flash floods, rock collapses, salting and too many tourists, these early records of the tombs are now precious historical documents. The Franco-Tuscan Expedition itself contributed to the damage – (or perhaps saved paintings from damage?) – since Champollion and Rosellini each removed an area of plaster with spectacular paintings from the tomb of Seti I, and these are now in the Louvre in Paris and the Archaeological Museum at Florence.

The expedition's letters were carried to and from Cairo by a courier who travelled on foot, and on at least one occasion when he failed to return, a second courier was sent to find him, dead or alive. On 2 April Champollion closed the last of a batch of letters since the courier was to leave the next morning. This letter to Jacques-Joseph carried the news that in the evening they were planning a special meal as a belated celebration of his daughter Zoraïde's fourth birthday, which they had been unable to celebrate on the correct day, 1 March, because at that time they were approaching the first cataract and few supplies were left. For this celebration meal they had at long last secured a crocodile – since leaving Alexandria the previous September, the expedition had been trying desperately to shoot or acquire a fresh crocodile to eat, the meat supposedly being a delicacy. Before the courier departed, a postscript was added to the letter: 'Our dish of crocodile turned during the night – the flesh has become green and stinking. What a misfortune!'

Although at times ill and overworked, Champollion was extremely happy during his months of adventure and exploration in Egypt. At last he felt he was fulfilling his destiny, and nowhere more so than at Thebes – he was therefore largely unaware that in Europe, and especially in Paris, controversy was still raging over his decipherment of hieroglyphs, fuelled particularly by the innuendo and slander of Jomard, Young and Klaproth. The sniping and backbiting continued, but Jacques-Joseph did not dwell on these matters in his letters, probably deliberately, which at times gave him little to say, leaving Champollion disappointed: 'I find your letters rather short. Remember that I am a thousand leagues from you.' Jacques-Joseph now informed his brother that Young had sent a letter to several academics in Paris, including their mutual friend the astronomer Arago, berating them for making too much of Champollion's discoveries. Completely confident of his own achievements, it was with a mixture of exasperation and pity that Champollion retorted:

> Is the poor Dr. Young therefore incorrigible? Why stir up an old business already mummified? Thank M. Arago for the cudgels that he so valiantly took up for the honour of the *Franco-Pharaonic alphabet*. Whatever the Briton does – *it will remain ours*: and all of *old* England will learn from *young* France to spell the hieroglyphs by quite another method than 'that of Lancaster' [a jibe at both Young and Jomard]. More-over, the Doctor still discusses the alphabet, and I, thrown for six months in the midst of the monuments of Egypt, I am ter-rified by what I *read* there, more fluently still than I dared imagine. I have extremely embarrassing results (*this between ourselves!*)

The letter continued with extremely vague hints that even Jacques-Joseph may have had difficulty understanding – Champollion was too frightened of the consequences to admit that the records relating to early pharaohs that he had so far amassed in Egypt utterly destroyed the Biblical chronology.

As the days of April passed into May, June and July, with the increasing heat of the Egyptian summer making conditions of work intolerable, all the expedition members were suffering – and not just from the heat. The explorer and physician Ricci had been stung on the arm by a scorpion and could no longer work. Although he stayed with the expedition, he never fully recovered and died in Florence in 1834. Others were suffering from exhaustion, not least Champollion, who was at the end of his physical strength and causing alarm among his companions. Insisting that he should be left alone to work in the tombs of the Valley of the Kings, he was found collapsed over his papers on several occasions. Some of his fellow travellers were tired of his endless obsession with the hieroglyphic texts, and L'Hôte complained about the hieroglyphs in a letter to his parents: 'We have been engulfed by them! A year of work, a year without interruption – with no day of rest, not one minute of respite.' Towards the end of July the artists Bertin, Lehoux and Duchesne all threatened to leave, but in the event it was only Duchesne who went, departing on a journey to Greece on 30 July, and taking some boxes of antiquities as far as Alexandria for Champollion.

Although exhausted, Champollion was still no less enthusiastic and optimistic, revealing his complete commitment in a letter to Jacques-Joseph at the beginning of July:

I am at last replying, my very dear friend, and a little late perhaps, to your three letters…But you must consider me as a man who has just come back to life: right up to the first days of June, I was an inhabitant of the tombs, where one is scarcely occupied with affairs of this world. Nevertheless, beneath those dark vaults, my heart lived and often crossed both Egypt and the Mediterranean, in order to reimmerse itself in the good memories of the banks of the Seine. These *family baths* refreshed my blood and intensified my heart…Do not forget to present at an appropriate moment my respects to M. de Sacy: I will be flattered if my results justify the kindness that he has shown for my work. I have had no reply to the two letters that I have written to M. Duke of Blacas…I would despair if they did not arrive and that he might deduce from my silence that I have forgotten all his goodness towards me:

such an oversight is not in my character – it is still less in my heart.

Having finished in the Valley of the Kings, the expedition moved its base to a house at neighbouring Qurna, which they called their 'château,' where they embarked on a routine of working from seven in the morning to midday and a further two hours in the afternoon, although the evenings could also be spent on reports, making duplicate copies of drawings and writing letters. Champollion began to study the inscriptions at the nearby site of Deir el-Bahari, only to be puzzled by what he found, because although he could read the names of two pharaohs, one name previously unknown to him had a feminine ending – yet it was associated with a portrait of a bearded pharaoh. Noticing that some names with the feminine ending had been deliberately erased, he deduced that one pharaoh had ruled as regent until the other came to power, a regency that must have been resented, leading to the erasure of the regent's name – but he was unable to explain the feminine ending. The site at Deir el-Bahari on the west bank of the Nile at Thebes, with its spectacular backdrop of limestone cliffs, was once the most dramatic of all the mortuary temples (where the cult of a dead pharaoh was celebrated), and Champollion was tantalizingly close to deducing that the pharaoh reigning as regent was actually a woman and that this was her mortuary temple – until then a concept alien to scholars. The female pharaoh was 〔hieroglyphs〕 (Maatkare Hatshepsut-Amun, usually known as Hatshepsut). The daughter of the Pharaoh Tuthmosis I, Hatshepsut came to power as regent for her young nephew Tuthmosis III following the reign of her half-brother Tuthmosis II. She took the unusual step of proclaiming herself pharaoh and probably ruled around 1498 to 1483 B.C., much of the time sharing power with her nephew. After she died her name was erased, probably more to reverse the blasphemy of a female pharaoh than through resentment on the part of her nephew.

After Deir el-Bahari, Champollion worked his way through the other sites of Thebes on the west bank of the Nile, including the complex of temples at Medinet Habu with its amazing sculptured reliefs. Here he correctly recognized and recorded the ancient Egyptian method of calculating the enemy casualties after a battle, describing a scene portrayed in one of the reliefs:

> The princes and the leaders of the Egyptian army conduct four columns of prisoners to the victorious king: scribes count and document the number of right hands and genital parts cut off the *Robou* [Asiatic people] dead on the field of battle. The inscription gives textually: 'Leading the prisoners into the presence of His Majesty; those are one thousand in number; hands cut off, three thousand; phallus, three thousand.' The Pharaoh, at whose feet these trophies are deposited, sitting peacefully on his chariot, whose horses are held by officers, addresses a speech to his warriors.

The temples and tombs of Thebes provided Champollion with ample evidence of his long-held opinion that Egyptian art developed first, uninfluenced by Classical Greek art, and that the apparently pure Greek art was, if anything, derived from Egyptian art. He presented his conclusions to Jacques-Joseph:

> Here is one of the thousand and one demonstrative proofs against the opinion of those who still remain obstinate in supposing that Egyptian art gained some perfection through the establishment of the Greeks in Egypt. I repeat it again: *Egyptian art only owes to itself* everything it has produced that is great, pure and beautiful...Old Egypt taught the arts to Greece, it gave them the most sublime development, but without Egypt, Greece would probably never have become the Classical land of the fine arts. Here is my complete profession of faith on this great question. I write these lines nearly oppo-

site reliefs that the Egyptians made with the most elegant skill and workmanship 1,700 years before the Christian era. What were the Greeks doing then?

In a letter to his brother in early July, Champollion gave his revised plans for the rest of the Egypt expedition – on 1 August he intended crossing the Nile to the east bank to study Luxor and Karnak, then on 1 September they would start on the journey home, stopping only at Dendera and Abydos on the way and arriving at Alexandria in the last days of September. A detour to Rome that he had been considering in order to complete his obelisk project was now abandoned, due to the death of his supporter Pope Leo XII in February, and so he urged his brother to ensure that a good boat was waiting at Alexandria for the expedition. Extremely fearful of spending the winter in Paris, he also gave instructions on the type of apartment that his wife Rosine should now find for them both, because during his stay in Egypt Rosine had been living with Jacques-Joseph's family in lodgings which went with his new post in the National Library that he had taken up the year before. Champollion urged that it should be 'especially a *warm* apartment, I have need of it in order to conveniently spend the harsh winter that awaits my return: it makes me shiver in advance.'

Not admitting in his letters that he was utterly exhausted, Champollion moved with the remaining members of the expedition across the Nile at the beginning of August, where their recording of the magnificent Luxor temple and its two obelisks was rapidly completed. In the letters that passed between the two brothers at this time there was much discussion about taking an obelisk to Paris, with Champollion in favour of one of the two obelisks outside this temple. He wrote enthusiastically:

I am returning again to the idea that, if the government wants an *obelisk* at Paris, it is a matter of national honour to have one of those at Luxor (that on the right when entering), a

monolith of the greatest beauty and seventy feet in height...
of an exquisite workmanship and astonishingly preserved.
Insist on that, and find a minister who might wish to immor-
talize his name by decorating Paris with such a marvel:
300,000 francs would do the job. Let it be considered seriously. If they wish to undertake it, let them send on site an
architect or a *practical* engineer (*but not a savant!*), his pockets
full of money, and the obelisk will march...I possess exact
copies of these two beautiful monoliths. I have taken them
with an extreme care, while correcting the errors of the
engraving of the *Commission* and completing them by the
excavations that we have done right to the bases of the obelisks. Unfortunately, it is impossible to record the end of the
east face of the right-hand obelisk and the west face of the left-
hand obelisk: for that it would have been necessary to demolish some earth houses and make homeless several poor
families of fellahin.

At the same time Champollion sent a letter to Drovetti, the French
Consul in Alexandria, urging him to influence the French govern-
ment to choose the Luxor obelisk rather than one at Alexandria.
Although he was not closely involved in the subsequent negotia-
tions, Champollion's chosen obelisk was eventually transported to
Paris and erected in the Place de la Concorde in October 1836.

Next to be tackled was Karnak, the enormous complex of religious buildings that covered around 250 acres of Thebes and which
had so enthralled the expedition members on their first visit the previous year. Their headquarters was the small temple of Opet, close
to the avenue of sphinxes leading to the Luxor temple. Considered
as a personification of motherhood and protector in childbirth,
Opet was normally portrayed as a female hippopotamus, and one of
the two main festivals at Thebes was the annual Opet Festival, when
a ritual procession of divine images was transported in ceremonial
boats along the avenue of sphinxes from Karnak to the Luxor

temple. The other main festival was the Beautiful Festival of the Valley, when cult statues were taken from Karnak across the Nile to the west bank.

After six months spent at Thebes, the expedition set sail for Dendera in the evening of 4 September 1829, visiting that temple once again the next day in order for Champollion to check the empty cartouche belonging to the zodiac: 'I wanted to reassure myself, *with my eyes and hands*, that the cartouches of the lateral inscriptions of the circular zodiac are really *empty* and were never carved: that is indubitable, and the famous *autocrator* is rather the work of our friend Jomard.' On leaving Dendera, they met up with two couriers bringing them letters, one from Jacques-Joseph admitting that in February Champollion had failed yet again to be elected as a member of the Academy of Inscriptions and Literature, due to the opposition of Jomard and his cronies. Even though he had been Secretary to the Academy's Perpetual Secretary Dacier for many years, Jacques-Joseph himself still only counted as a 'corresponding member' based in Grenoble.

The boats of the expedition made their way rapidly downstream, at a time when the Nile inundation was particularly severe – the flooding was regarded as a magnificent spectacle by Champollion, although at the same time he was distressed by the plight of the peasants, whose harvests and fields had been ruined. The intended visit to Abydos had to be abandoned because of the flood, which was probably not too bitter a blow, considering the fragile state of Champollion's health: he was by now looking forward to a brief stop at Cairo, then on to Alexandria and home. Against all advice, and being very unwell, the naturalist Raddi now decided to set out into the Delta by himself, his assistant Galastri having returned to Italy some months previously because of illness. It is assumed that Raddi became lost in the Delta – he was never seen again.

At Cairo, Champollion was upset to hear that his old adversary and colleague Thomas Young had died a few months earlier. They had last met in Paris on the eve of Champollion's departure for

Egypt, a country that Young never visited. Young had gone on to Geneva from Paris, and on his return to his Park Square house in London in the autumn of 1828, he had settled down to a rather pleasant life, enjoying his normal good health and commenting soon after his return: 'As for myself I am perfectly content with the life I lead: walking on business of routine every day from eleven to two: the rest of the day sitting over my hieroglyphics or my mathematics, and conversing in my library with people beyond the Alps or the Mediterranean.'

Young continued to work on the demotic (his enchorial) dictionary to be appended to Tattam's book on Coptic grammar, writing to his old friend and former pupil Hudson Gurney in mid-December (when Champollion was heading into Nubia from Philae) that:

> I have just finished the fair copy of my little Egyptian Dictionary, except that I must copy it all over again as the lithography goes on, which will be the work of two or three months for the fingers and the eyes, but little or nothing for the head. It contains little or nothing striking; but it preserves from oblivion all that I have made out in the running hand, which is no where methodically recorded. It makes about a hundred pages.

Writing to Young from Naples a month earlier, Gell had said:

> I wish you had sent me to Egypt with Champollion, who offered to take me, but I had no money. I have no doubt I should have done something, as I think I take views and plans quicker than my neighbours, and have more patience in working out the hieroglyphics...I am glad you are friends again, only publish your Enchorial Dictionary soon, that some ugly German may not do it before you, for I dare say some dozens of them are plodding over it at the present moment and do nothing else.

From February 1829, while Champollion was returning to Philae from Nubia, Young appeared to be suffering from asthma attacks and weakness, and by April his lungs and heart were severely affected. Confined to his bed, he declared that he had finished all his research except for his dictionary, whose printing he continued to supervise, although he was by now in a very weak state, only being able to hold a pencil and not a pen. He said that the compilation of the dictionary was a great amusement to him and that 'if this disease cut short his days, it was satisfaction to him never to have been idle through life.' Finally, on 10 May 1829, when Champollion was in the tombs of the Valley of the Kings, Young died at the age of fifty-five from 'ossification of the aorta' and was buried with his wife's family in the vault of St. Giles the Abbot Church in the village of Farnborough in Kent. His wife was anxious for a memorial to be placed in Westminster Abbey, which took the form of a medallion sculpture with an inscription written by Gurney. A few other memorials of Young also survive today, including a plaque on the house where he was born at Milverton in Somerset and another on his house at Welbeck Street in London. In Shire Hall of the County and Crown Courts in Taunton, Somerset, is a marble bust of Young, beneath which a shorter version of the Gurney inscription reads:

> Thomas Young, M.D. Fellow and Foreign Secretary of the Royal Society, Member of the National Institute of France, unsurpassed in the extent and accuracy of his knowledge. Second only to Newton in the keenness of his insight into the causes of phenomena especially those of physical optics. The first to discover Egyptian hieroglyphs. Author of lectures of natural philosophy. One of the most Classical specimens of scientific literature. Endeared to his friends by his domestic virtues. Honored by the world for his renowned acquirements. He died in the hopes of the resurrection of the just. Born at Milverton in Somersetshire June 13th 1773. Died in

Park Square London May 10th 1829 in the 56th year of his
age.

Young's *Rudiments of an Enchorial Dictionary in the Ancient Enchorial Character; containing all the words of which the sense has been
ascertained* was published the year after his death through the
efforts of Tattam and Gurney, with some of it compiled from his
rough notes; it was accompanied by a report of his life and a list of
his published works. His book included comments on Egyptian
chronology using mainly astronomical evidence, followed by sections on demotic numbers and the names of months and dates, in
which many of the papyri traced for him by Champollion were used
and acknowledged. The concluding and most extensive part of this
work was a dictionary of demotic words and phrases with English
translations. Despite his failure to make the necessary break-
throughs in hieroglyphic decipherment, Young became the first
scholar to make real advances in the study of demotic – the language and writing of ancient Egypt from the mid-seventh century
B.C. Although confused about demotic and hieratic in his early
studies, Thomas Young should probably be regarded as the true
decipherer of demotic, and it is a pity that the controversy over his
part in the decipherment of hieroglyphs has tended to overshadow
this achievement.

Towards the end of September, four months after Young had
died, the remaining members of Champollion's expedition left Cairo
for Alexandria, anxious to be on time for the arrival of the ship from
France. At Alexandria they discovered that the boxes of antiquities
brought there by Duchesne when he left the expedition a few
months earlier had not been placed in the safe keeping of the new
French Consul who had replaced Drovetti, but had been left in the
hands of a merchant, with the result that many items were missing.
Now began a highly frustrating and wasteful period, because the
promised vessel from France had not even left Toulon. In the meantime Champollion had several meetings with Mehemet Ali, the

ruler of Egypt, who asked him to write a summary of the ancient history of the country. Fulfilling this request, Champollion's report revealed for the first time his latest thoughts on the antiquity of Egypt, which he dared to state went back 'six thousand years before Islam' – at least a thousand years earlier than the date of the creation of the world according to Christian theology.

Mehemet Ali also received a second report that he had not asked for, on what was for Champollion an urgent problem – the preservation of the monuments of Egypt and Nubia. In this report he tactfully but sorrowfully stated that many travellers and scholars 'bitterly deplored the destruction of a host of ancient monuments, totally demolished in the last few years, without there remaining the least trace of them. One knows well that these barbaric demolitions have been done against the enlightened views and well-known good intentions of His Majesty.' Champollion went on to list those monuments that had been recently destroyed and those that should now be assured of preservation at all costs, finishing with the advice that although archaeological excavations should not be stopped, as the information they produced was too valuable, the excesses of excavators should be curbed:

> In summary, the well-agreed interest of science demands, not that the excavations are interrupted, because science acquires new certainties and unexpected enlightenment each day through this work, but that one submits the excavators to such a control that the preservation of the tombs discovered today and in the future might be fully assured and well guaranteed against the attacks of ignorance or blind greed.

Rosellini and the remaining members of the Tuscan expedition, tired of waiting for the promised French transport, sailed back directly to Livorno on a merchant ship, so that only Champollion, Cherubini, L'Hôte, Bertin and Lehoux were left in Alexandria. The three French artists decided to stay on in Egypt as they had been requested to paint portraits and theatre scenery, leaving Champollion alone with the loyal Cherubini to sail back to Toulon at the beginning of December on the *Astrolabe*. It was on 23 December 1829, Champollion's thirty-ninth birthday, that they finally arrived in Toulon harbour, to face one month of quarantine, which was regarded as a necessity because bubonic plague was a constant threat in Egypt. Most of the dreary days in quarantine in the worst winter in living memory were spent in the dirty, bare quarantine station, with a smoky oven for heat, and they slept on board the unheated ship.

Released from quarantine towards the end of January 1830, Champollion was dreadfully worried about returning to Paris, knowing that his health would suffer even more there. His intention was to stay in the cold but drier conditions of the south, visiting friends and seeing Egyptian antiquities until the end of February. 'What a demon of winter the heavens are thus sending us this year,' he wrote to Jacques-Joseph, 'I am suffering greatly from it, and strongly fear finding gout on arriving in the foggy atmosphere of Paris.' His reluctance to return to Paris was compounded by hearing that opposition to him had increased during his visit to Egypt, and he was even being accused of falsifying evidence to fit his system of decipherment. Writing to Rosellini, Champollion declared that his first concern would be to finish his hieroglyphic grammar, which 'will appear at the end of this year: it is the indispensable preface to our voyage. It will not convert, however, those who fight my system and deprecate my work, because these gentlemen do not wish to be converted and are uniquely dishonest...I spit on them.'

At two o'clock in the morning of 4 March Champollion returned to Paris, where a violent attack of gout immediately confined him to his new apartment on the second floor of 4 rue Favart, just minutes from where Jacques-Joseph lived and close to the Louvre. His low-key arrival in Paris was in direct contrast to the return of the members of the Tuscan Commission, who were enthusiastically received and showered with acclaim and honours in Italy. Grand Duke Leopold II was anxious to have all the results published as quickly as possible, and so at the end of April Rosellini began to press Champollion for his contribution, urging him to come to Pisa to stay with him and his wife: 'Take your papers and come here...I would keep you company with my wife, and we will be able to spend the whole summer together and also the autumn...Mme Rosine and the little sweetheart could very well come with you.'

In Paris the situation was not so simple, and it was increasingly difficult for Champollion to concentrate on any one task. The antiquities from Egypt had not long arrived at the Louvre, and with the purchase of Drovetti's collection just before the Egypt trip a new catalogue for the whole department was urgently required. In the midst of continual calls on his attention, he was trying to work on his hieroglyphic grammar before embarking on the preparation of the major report on the results of the Egypt expedition, as well as cope with his still precarious health. In March he had again been excluded from the Academy, to the absolute disbelief of many academics across Europe and to the discredit of the Institute of France. Stung into action, the Academy therefore undertook a mass election of new members, and Champollion was finally admitted on 7 May 1830.

Nine days later Joseph Fourier died at the age of sixty-two, memories of Egypt very much in his thoughts, as Champollion had only seen him a few days before to talk about his own expedition. Having lost his post as Prefect of the Isère in 1815 in the wake of Napoleon's return from exile, Fourier had only remained for a few weeks in his new post of Prefect of the Rhône. Returning to Paris, he first worked in the statistical office and then for the Academy of Sciences, becoming the Perpetual Secretary of its mathematical section in August 1822, and over the last few years he had been greatly interested in Champollion's progress. Pointing to the Pantheon, the burial place of the illustrious of France, on the eve of Champollion's departure for Egypt, Fourier had encouraged him with the words: 'It is *Egypt* which, one day, will place you in that sanctuary.'

In the second half of 1830 several European countries were experiencing revolutions, and once again France was no exception, with liberals increasingly opposed to the ultra-conservative government of Charles X, which had become repressive. The Royalists dissolved the newly elected Chamber and in effect abolished the constitution on 25 July, causing the people to take up arms in what became known as the revolution of the Trois Glorieuses (Three Glorious Days) from 27 July. Connected at that time to the Tuileries palace, the Louvre was being used as a base by the military forces protecting the King, and so on 29 July thousands of armed citizens broke into the Charles X Museum at the Louvre, but most of them were more intent on looting than revolution. They stole many statuettes, figurines, amulets, papyri and objects of gold, silver and precious stones from Champollion's Egyptian galleries, including ones he had just brought back from Egypt and items belonging to the Salt and Drovetti collections, which were never seen again – a tragic loss both for Champollion and for Egyptology.

Charles X fled to England, and Louis-Philippe, Duke of Orléans, was proclaimed King of the French, rather than King of France; the political situation in the country was still in a simmering state as a great number of people were dissatisfied with the prospect of the new monarchy. Many of the nobility gave up prominent jobs, and the Duke of Blacas prepared to go into exile, as he had been too closely connected with Charles X. Only weeks after the death of Fourier, Champollion was devastated at having to bid farewell to Blacas, who had been his friend and supporter for so long, and to Lenormant he confessed: 'This separation has aged me by years.'

When Champollion returned from Egypt, Blacas had been unsuccessful in persuading King Charles X to give him a token of gratitude in the form of a personal chair of Egyptology, but Louis-Philippe was much more sympathetic to Champollion, who was able to call on the new King to discuss matters of interest, including

the projected transport of the obelisk from Luxor to Paris. After all
the turmoil of the revolution, it was only in late September that
Champollion was able to resume his normal course of work, and
having delayed writing to Rosellini for so long, he finally outlined
his plans to him: in October and November he wanted to sort out
the remaining problems of his hieroglyphic grammar, which would
begin to be printed in the next two months, and at the same time he
intended to compile the great joint work on the Egypt expedition.

Champollion's tardy letter reached Rosellini just as he was on
the point of announcing the proposed publication in Italy, which he
had anticipated publishing periodically from the following year,
and his reply of early October was full of anger at this new delay.
Revolution was still bubbling under the surface in France, and in
December the Louvre, once again being used as a military headquarters, was briefly threatened by attackers. Along with the
worries brought about by this political turbulence, Champollion
was also besieged at home and in the Louvre by admiring visitors
from France and abroad, and was weighed down by an increasing
workload, such as the need for an overhaul of the museum, with the
result that his personal research could only be carried out at night,
to the continued detriment of his health.

One project in which Champollion was involved, in collaboration with the astronomer Jean-Baptiste Biot, was the analysis of
the notes and drawings from Egypt concerning the calendar, the
seasons, the agricultural year and astronomy. Reports on these subjects were read in March and April 1831 to the Academy of
Sciences and the Academy of Inscriptions, creating a profound
impression on the audiences, because it was shown that absolute
dates could be determined using the evidence of ancient astronomical events, such as eclipses and solstices. At the end of the most
important meeting on 18 March, Champollion was presented with
a royal decree appointing him Professor of the College of France,
where he had been a student only thirteen years previously.

The harshness of the previous winter, with the weeks spent in dreadful conditions in quarantine, had affected his lungs and throat, so that for the last few months he had experienced difficulty in speaking for long periods and was more and more absent from the Louvre. Increasingly he relied on the help of his assistant Dubois and on the fortunate appearance of a student who had come some months earlier to study under his direction. Francesco Salvolini, now twenty-two years of age, had studied Oriental languages at Bologna and was highly recommended by Champollion's friend Gazzera at Turin, who considered that Salvolini showed more promise than Rosellini. The student was very willing to assist in any way at the Louvre, run messages to and from the museum, and return to Champollion's apartment to help with the hieroglyphic grammar and the preparation of the university course, giving him a unique opportunity to gradually acquire a detailed knowledge of Champollion's latest ideas on hieroglyphs and Egypt in general.

Champollion now prepared to teach his first university course, and the advance publicity announced that: 'M. Champollion will explain the principles of *Egyptian-Coptic grammar*, and will develop the entire system of *sacred writing*, while making known all the grammatical forms in common use in the *hieroglyphic and hieratic* texts.' The inaugural lecture was intended to become the preface to the forthcoming publication on the grammar, but because of his ill-health it was postponed for a few weeks. On 10 May 1831 Champollion finally presented his inaugural lecture as Professor at the College of France, in which he gave a summary of the developing knowledge about the writing of ancient Egypt. The study of ancient languages, he maintained, was based on a combination of philology and archaeology: 'It is principally in the domain of these two sciences brought together, archaeology and philology, indispensable auxiliaries of history, that belong, through their own *essence*, if one can express it thus, the monuments of old Egypt, the main object of the course which opens today.'

The response to the first lecture was extremely heartening, with many scholars from across Europe making the effort to attend this momentous occasion, but the strain on Champollion was severe. He suffered an acute worsening of his bronchial problems, which confined him to his apartment, and the next university class intended for two days' time was postponed for two weeks. He wrote to Leopold II in Tuscany to ask him to allow Rosellini, who was still very irritated, to come to Paris to sort out the details of their joint publication. At the university only two more lectures were given by Champollion before his difficulty in speaking forced him to abandon the course for that term. He now began to be affected by acute and repeated attacks of gout, so that everyone around him advised him to leave Paris, particularly with the outbreak there of an influenza epidemic – advice he was unable to heed as Rosellini was expected any day.

In mid-July Rosellini arrived in the capital, in the midst of oppressive summer heat and riots, and was deeply concerned to find Champollion so ill. Towards the end of July, when Champollion had partly recovered, they both worked on the material from Egypt, and Rosellini was informed of all his latest ideas and discoveries. Jacques-Joseph, always trying to protect his brother, who was eager to share his latest discoveries, ensured that the arrangements for the joint publication were enshrined in a legal contract. Champollion wrote to King Louis-Philippe to ask for permission to dedicate the work to him, a contrast to Champollion's angry attempt to prevent his first book being dedicated to King Louis XVIII seventeen years earlier. A private meeting was arranged with Louis-Philippe in mid-August, in which he accepted the dedication, and a few days later, on 21 August, Champollion quietly left for his home town of Figeac, feeling that he could no longer breathe the air in Paris.

Arriving in Figeac four days afterwards, he went to the family home in the street by then renamed Rue Champollion in his honour. His two sisters, Marie and Thérèse, were overjoyed to see him, and his visit aroused the immediate interest of many in the town. Already feeling stronger now that he was out of Paris, he began straightaway to write out a good version of his grammar for publication, just leaving the house briefly in the middle of each day for a short walk. His sisters protected him from the unwanted attentions of visitors and did everything to ensure his stay was comfortable and restful. Champollion was very anxious to complete this publication, because with threats of war, revolution and the rampant cholera epidemic raging across Europe, he was so fearful of the future. He still worried about publishing too soon, though, confessing several times to Lenormant that if he was certain of several more years to live, he would not publish now. Intending to return to Paris in November, Champollion urged his brother to arrange for the printing of his book. Jacques-Joseph had by now settled the details of the legal contract with Rosellini, who left Paris for Italy in early September with a signed copy.

His health much improved in Figeac, Champollion was reluctant to return to Paris, just as he had been during his quarantine at Toulon twenty months earlier. Writing to Jacques-Joseph, he advised him that he would only return when it was necessary to begin his university course, and that he did not wish to start at eight o'clock in the morning, as that was too harsh in winter. This was the first time since his voyage to Egypt that he was able to work on his research in peace, but it was not to last. Jacques-Joseph insisted that he return immediately, because the Louvre was anxious for his presence, de Sacy wanted him back at the College of France, even though he knew that the grammar was to be dedicated to him, and the minister of the navy wanted to discuss the transport of the obelisk from Luxor. 'Death lies in wait for me at Babel,' Champollion mournfully said to a friend in Figeac, and to his brother he wrote that he only needed to be left alone for a little longer to complete his hieroglyphic grammar: 'Only one month more – and my 500 pages would be finished. But it is necessary to give up and be pleased with what is possible.'

Taking the diligence to Paris on 28 November, Champollion intended arriving there three days later in order to resume his course on 2 December, but in Lyons rioting had erupted, and the delay in reaching Paris during the cold weather affected his health. It was only on Monday 5 December that he resumed his course at the College of France, before an audience that was captivated by his clarity of expression and his incredible enthusiasm. He managed to give one more lecture, but had barely started the next on 9 December when he collapsed. Four days later he suffered a stroke, which left him partially paralyzed, and although within a few days he was able to leave his bed, he had great trouble writing. Extremely worried, he entrusted his brother with his grammar manuscript and notes.

Rosellini meanwhile was hoping for the publication of their first volume in January 1832 and urged Champollion yet again to come to stay with him and his wife: 'I would put myself in charge of all the cares of your life – with me and my wife you will surely feel at home.' Gradually improving, Champollion thought that he would soon be able to resume his work, but fell into bouts of depression. Much of the time the student Salvolini kept him company, but Jacques-Joseph was suspicious of his motives, having heard a rumour that he was a spy who had been sent by the King of Sardinia-Piedmont.

On 23 December 1831, on his forty-first birthday, Champollion insisted on being taken back to see his room at 28 rue Mazarine, where he remained for a while, deeply moved, announcing: 'This is where my science was born, and we form an inseparable entity – we are one!' At the same time he visited Dacier, who had been bedridden for some years and who often called for Champollion to visit him. On 11 January 1832, the astronomer Biot came to see Champollion, and they both enthusiastically discussed plans for further innovative research using the evidence of astronomical events for dating. Resuming the next day, they began to map out the work when Champollion collapsed with a cry of pain. Urgently called to assist, doctors found him suffering from acute gout and paralysis of movement, in addition to which he could barely talk. These symptoms gradually eased, and towards the end of January some improvement was noticeable, although Champollion was in a distressed state, crying out: 'My God, two more years still – why not!' and pointing to his head declared: 'Too soon – there are so many things to do in there.'

His illness dragged on for several weeks, and by the end of February he was drifting in and out of consciousness. Late in the evening of 3 March, Champollion suddenly became alert and regained his speech in a remarkable way, but sensing that the end was near, Jacques-Joseph called for a priest to give his brother the last rites. A few friends and family members came to say goodbye, including his little daughter Zoraïde, just turned eight years of age. His nephews had maintained a vigil over their uncle for more than two months, and now Champollion asked them to bring from his office some of the few objects that he had kept from Egypt: his Arabic costume and sandals, and his notebooks. Two years and two hours after his return to Paris from Egypt, towards four o'clock in the morning of 4 March 1832, Jean-François Champollion died.

The news of his death sent shock waves throughout Paris – many people who knew of his illness had not suspected that it was so serious since the family had withheld information from the newspapers. Because he had changed in appearance so utterly, few people were allowed to see his body, and towards eleven in the morning on 6 March Champollion was taken to the nearby church of Saint-Roch, where he had learned Coptic with the priest nearly twenty-five years before. From here the massive funeral cortège set off eastwards, in the midst of the milling crowds celebrating Mardi Gras, to the Père Lachaise cemetery which had been established by Napoleon outside the walls of Paris and where Napoleon had asked in vain to be buried.

The funeral address was given by M. Walckenaër, President of the Academy of Inscriptions and Literature, who lamented that death had taken away a genius so young, ending with the words:

He had just finished his *Egyptian Grammar* and had just announced its printing, when he was suddenly taken from a family who loved him dearly, from our Academy where he had as many friends as colleagues, from France who counted him amongst the number of its illustrious men, from learned Europe who had already inscribed his name in its literary splendours. This name will never perish; but the vivid light that M. Champollion spread over the land and monuments of ancient Egypt went out at the very moment when it was shining at its most brilliant, and the shadows that we were hoping to see disperse in its light leave us with sorrows which, perhaps for a long time still, will be shared by posterity. The bereavement of a single family becomes a general bereavement for all those who appreciate literature and who are interested in its progress.

# CHAPTER TEN

# (... Who Gave Words and Script)

Jacques-Joseph never did see Egypt. He always regretted being denied a place on Napoleon's expedition in 1798 and no other opportunity arose as his career unfolded. Yet he was to play a crucial role in the development of Egyptology by editing and publishing the work that his brother had been unable to complete. When Champollion died, Jacques-Joseph was devastated – he was now fifty-four years old, and, after all the experiences they had shared, the premature death of his younger brother affected him deeply.

In an age of uncertain medicine, the doctors who had been unable to save Champollion's life were equally at a loss to determine what exactly had killed him. Apart from recurring attacks of gout, the reports of the illness leading up to his death are conflicting, but he may well have been suffering from a major malady such as tuberculosis, heart disease or diabetes. What is more certain is that he finally died of a stroke. During his all-too-short years, Champollion had lived through the French Revolution, the rise and fall of Napoleon and the reign of three kings of France. He had travelled through Egypt and Italy, taught in schools and universities, organized the Egyptian collections in the Louvre Museum, and had done so much research on so many aspects of ancient Egypt that much of his important work was still unpublished. Realizing the dream that had obsessed him for so long, he left behind him one great achievement as his memorial – the decipherment of hieroglyphs.

Jacques-Joseph's first task was to persuade the government to purchase all his brother's unpublished papers in order to ensure their safekeeping and to provide for Champollion's widow Rosine and daughter Zoraïde. Included in these papers was a massive amount of material relating to the Egypt expedition, as well as the manuscripts of his hieroglyphic grammar and dictionary, his Coptic grammar and dictionary, and numerous notes on topics such as ancient Egyptian religion and the chronology of the country. Jacques-Joseph was shocked to realize that some of the manuscripts were missing, including work Champollion had done on the Egyptian numerical system and over half of his hieroglyphic dictionary. He let friends know what was missing, straightaway suspecting Salvolini, who had been given free access to Champollion's office during his illness. Salvolini steadfastly and indignantly refused to admit that he was guilty, but now began to publish pioneering articles on hieroglyphs that met with much acclaim.

At first the government stated it could not afford to purchase Champollion's papers, but the press pursued a vigorous campaign that resulted in the setting up of a commission to consider the proposal, and all the papers were temporarily moved to the safety of the National Library. The report of the commission considered that the Coptic grammar was not worth publishing but recommended the purchase of the rest, observing that the gathering of material during the Egypt expedition might have been partly responsible for his death:

One has trouble in conceiving that a single man during quite a short stay in Egypt could have composed without help this enormous mass of descriptions and notes, in which one finds, on nearly every page, a more or less great number of hieroglyphic inscriptions. One has trouble in defending the terrible idea that such vast efforts and a truly colossal work must have contributed to shortening a life so precious for science.

In April 1833, just over one year after Champollion's death, the government finally voted to purchase his manuscripts for a sum of 50,000 francs and award a pension of 3,000 francs to Rosine – the manuscripts now comprise eighty-eight volumes in the National Library at Paris.

Silvestre de Sacy, one of the members of the commission who decided to purchase the papers, was by now Perpetual Secretary of the Academy of Inscriptions and Literature, having recently succeeded Dacier. At a public meeting of the Academy in August 1833 he read a lengthy eulogy of Champollion, noting the danger that his death would once again plunge ancient Egypt into darkness and obscurity:

> It was not only a distinguished savant that the literary world lost; with him all science and all the arts of ancient Egypt seemed to be swallowed up in his tomb and to return to the domain of shadows and death; and the daylight, which had begun to shine on the monuments of Thebes and Memphis and had begun to make them come out of their ruins, vanished like those fantastic lakes that a light haze creates in the desert and which disappear before the thirsty traveller, at the moment when he thinks he has reached them and will quench the thirst which burns him. Let us hope that it will not be thus; that the works of the ingenious and indefatigable Champollion will yet give birth after him to heirs of his genius, who will cultivate the field that he was the first to clear.

After having described the progress of Champollion's life and his work, de Sacy finished his eulogy by reminding his audience that:

> Apart from the decoration of the Legion of Honour, M. Champollion had received that of the Order of Merit of Tuscany. The Academies of Göttingen, Petersburg, Turin, Stockholm, the Royal Asiatic and Literary Societies of London, and several

other learned societies, national or foreign, had hastened to vie in making him one of their number of associates. He has been replaced in the Academy of Inscriptions and Literature by M. Eugène Burnouf. The chair created for him at the Royal College of France is still vacant.

Jacques-Joseph now embarked on preparing his brother's work for publication, the first and easiest task being to publish the *Grammar*, of which he had most of the manuscript, although part was missing, presumed stolen by Salvolini. During his last stay at Figeac, Champollion had made a good copy of over half the manuscript, which became the first volume published by Jacques-Joseph, who dedicated it to de Sacy and presented a copy to him on what would have been Champollion's forty-fifth birthday – 23 December 1835. The *Grammaire égyptienne, ou principes généraux de l'écriture sacrée égyptienne appliquée à la représentation de la langue parlée* (Egyptian grammar, or general principles of sacred Egyptian writing applied to the representation of the spoken language) was published in parts over a period of five years from 1836. In the preface Jacques-Joseph explained how much effort Champollion had put into this, his last work, and that his brother had entrusted the manuscript to him during his final illness with the words: 'Hold it carefully, I hope that it will be my calling card for posterity.'

By the time the *Grammar* was published, Salvolini was increasingly held in contempt because it was beginning to be realized that at least some of his work was not his own. In August 1833 at the public meeting of the Academy, de Sacy had called upon the person or persons who had the manuscripts of Champollion to give them up, but Salvolini only declared his distress at the terrible events. By a strange sequence of coincidences, the full extent of his plagiarism was unmasked only a few years later due to his early death in Paris in February 1838, at the age of twenty-eight. Rather than asking for the return of his papers, his family in Italy persuaded an old friend in Paris to dispose of them, but nobody wanted to buy them, and it was only when Charles Lenormant was consulted for advice that the manuscripts of Champollion were recognized.

With the return of the stolen manuscripts, Jacques-Joseph wrote a short report for friends on how the missing papers had been retrieved and then set to work on the hieroglyphic dictionary, a task requiring a great deal more effort, as the manuscript material was nowhere near ready for publication. Not knowing how to arrange the words, he finally grouped them according to subject (such as birds and animals), although a more alphabetical approach to arranging entries is used today. The *Dictionnaire égyptien en écriture hiéroglyphique* (The Egyptian dictionary of hieroglyphic writing) was published in parts from 1841 to 1844, but was extremely difficult to use, and researchers complained about the lack of an index. Jacques-Joseph's devotion to the publication of his brother's work at the inevitable expense of his own research is nonetheless admirable, and even though the posthumous publications of the *Grammar* and *Dictionary* included errors that Champollion himself would probably have eradicated, they provided a huge impetus for the study of hieroglyphs, as a new wave of Egyptologists emerged throughout Europe – particularly in Germany, Britain and France.

The publication of the results of the French expedition to Egypt and Nubia was more complicated, as Jacques-Joseph refused to liaise with the Tuscan expedition in an attempt to protect Champollion's interests. Rosellini therefore went ahead and published several volumes in Italy, but never completed the work as in 1843 he died at the age of forty-two, just one year older than Champollion. From the records of the French expedition, the incredible drawings of all the sites with short explanations were prepared first of all – four volumes with more than 500 plates, published over ten years from 1835. In order to fill in some gaps, Nestor L'Hôte was sent to Egypt by the government in 1838, where he reproached himself bitterly for not having appreciated Champollion more while he was alive, particularly during the expedition in Egypt. His feeling of desolation was made worse by seeing how much Egypt had changed over the last ten years, even in the Valley of the Kings. Here he set up his bed in the tomb of Ramesses IV, where Champollion had once slept, and took comfort remembering the time spent there with him and in conjuring up his image.

Yet another revolution erupted in France in 1848, overthrowing King Louis-Philippe and establishing the Second Republic under the presidency of Josephine's grandson Louis-Napoleon (who became Emperor Napoleon III in 1852). Losing his job at the National Library, Jacques-Joseph was forced to abandon the publication of the French expedition report, and it was only in 1868, the year after his death, that work on the text was resumed. The last volume appeared in 1889 – sixty years after Champollion left Egypt. This final work was entrusted to the young Gaston Maspero, who would become the foremost Egyptologist of the later nineteenth century.

The tragic death of Champollion did not soften the attitude of some of his opponents, most notably Jomard and Klaproth, who continued to vilify his work. While many scholars would not accept the importance of Champollion's results, and published articles against him, others, such as Richard Lepsius in Germany, Samuel Birch in England, Samuel Hincks in Ireland, and Emmanuel de Rougé in France, acknowledged his vast achievements and took steps to advance the understanding of hieroglyphs. Lepsius taught himself hieroglyphs from the newly published *Grammar* and started to expand and correct Champollion's system of decipherment; one of his major advances was to recognize that there were not just uniliteral phonetic signs – consonants with one syllable – but phonetic signs with two and three syllables (biliterals and triliterals).

As keeper of the collections at Berlin's Egyptian Museum, Lepsius went on his second expedition to Egypt in 1866, exploring the Delta and Suez Canal regions. Tanis (modern San el-Hagar, seventy miles north-east of Cairo) was the most important site in the eastern Delta, with royal tombs dating from around 1000 B.C., but Champollion never had time to visit it. Here Lepsius studied a limestone stele that his expedition had discovered – a stele that represented the publication of a decree of priests in 238 B.C. during the reign of Ptolemy III. The decree was to reform the Egyptian calendar, and this stele was one of many set up in temples in Egypt. The priests had met at Canopus (the port later known as Aboukir), and so the stele found at Tanis became known as the Canopus Decree or the Tanis Tablet. Now in the Egyptian Museum at Cairo, its bilingual decree was, like the Rosetta Stone, written in three scripts and two languages, with thirty-seven lines of hieroglyphs and seventy-six lines of Greek on the front face, and seventy-four lines of demotic on one side – the latter was not even noticed at the time of discovery. These inscriptions provided the final verification that the decipherment system of Champollion and his successors was absolutely correct.

Voices were still being raised against Champollion in the latter half of the nineteenth century; Sir Peter Le Page Renouf, an Egyptologist and Orientalist from Guernsey, campaigned tirelessly on his behalf for decades, writing in 1863 that

> if so much remains to be done by his successors, how can it be true that Champollion 'read hieroglyphic texts with facility and certainty'? The fact is, he did read *some* texts with facility and certainty, and could not read others. Hieroglyphics, like all other texts, differ greatly as to facility of translation. Some are comparatively easy, others are extremely difficult. Some are as yet untranslatable.

The attacks persisted, and as President of the Society of Biblical Archaeology Renouf gave a speech over three decades later in 1896 in which he attempted to demolish the past and present arguments in support of Young and against Champollion, finishing with the words:

> Two undeniable facts remain after all that has been written: Champollion learnt nothing from Young, nor did anyone else.

It is only through Champollion and the method he employed that Egyptology has grown into the position which it now occupies. It is only by the strictest application of that method that Lepsius, Birch, and de Rougé were able to correct the errors and imperfections adhering to the system founded upon it, but in no way pertaining to its essence.

Gradually, silence descended on most of Champollion's opponents – those who were still inclined to attack him were reduced to carping that he had not sufficiently acknowledged Young's achievements.

The impact of the decipherment of hieroglyphs was almost unbelievable – in effect, it meant the discovery of a whole new civilization. Once translation of the ancient Egyptian texts written in hieroglyphs and hieratic began in earnest, an astonishing miscellany of information about ancient Egypt became accessible. A British Egyptologist, Francis Llewellyn Griffith, summed up the situation in 1922 when he wrote that 'Champollion turned bewildering investigation into brilliant and continuous decipherment,' and this process of decipherment was extended by the Egyptologists who came after him, including Griffith himself. What became available from all these translations was astounding in its quantity and variety: as well as texts written on papyrus, wooden boards and leather, there were notes scratched on pieces of broken pottery and stone (ostraca), painted and carved inscriptions all over the walls of temples and tombs, and writing on a huge variety of objects, from colossal statues of the pharaohs down to the linen bandages used to wrap their bodies during the mummification process. At last the expectations of the savants who had taken part in Napoleon's Egypt expedition were being fulfilled: it was they who had dreamed of unlocking the secrets of ancient Egypt, and now that dream was coming true. Even the words themselves, such as 🔲⚬𓏤 'annals'; 𓏌𓂝𓏏 'hammock'; ☰⚬𓏤 'green eye-shadow'; 𓏌𓂧𓏊 'beer'; 𓏌𓏏𓀏 'archer'; 🔲𓏏𓏊 'a lie'; and 𓏏×𓏤𓀏 'taxpayers,' showed the complexity of Egyptian civilization and the huge range of information that began to emerge as a result of Champollion's success.

It is the sheer volume and diversity of information about ancient Egypt in the hieroglyphic and hieratic texts that have made their decipherment so important. The survival of huge numbers of documents, including thousands upon thousands of papyri and ostraca, is due to the combination of a favourable climate and the attitudes of the ancient Egyptians themselves. They regarded their scribes as the most important people (the pharaoh was a scribe), who were writing not just for their contemporaries but for the generations that came after. This conscious attitude is seen in numerous texts which extolled the high status of scribes. Of these, a didactic text that has survived at least 3,000 years sets out to show that the written word outlasts everything else that can be made. It is now known as the 'Eulogy to Dead Authors':

> But, should you do these things, you are wise in writings.
> As for those scribes and sages
> from the time which came after the gods
> – those who would foresee what was to come, which
>     happened –
> their names endure for eternity,
> although they are gone, although they completed their
>     lifetimes and all their people are forgotten.
> They did not make pyramids of bronze,
> with stelae of iron.
> They recognized not how heirs last as children,
> with offspring pronouncing their names;
> they made for themselves heirs
> as writings and the teachings they made.
> They appointed for themselves the book as the lector-priest,
> the writing board as beloved-son,
> the teachings as their pyramids,
> the pen as their baby,

the stone surface as wife.
From the great to the small
are given to be his children:
the scribe, he is their head.
Doors and mansions were made: they have fallen,
their funerary priests leaving
while their stelae are covered with earth,
their chambers forgotten.
Yet their names are still pronounced over their rolls
which they made, from when they were.
How good is the memory of them and what they made –
for the bounds of eternity!
Be a scribe! Put it in your heart,
that your name shall exist like theirs!
The [papyrus] roll is more excellent than the carved stela,
than the enclosure which is established.
These act as chapels and pyramids
in the heart of him who pronounces their names.
Surely a name in mankind's mouth
is efficacious in the necropolis!
A man has perished: his corpse is dust,
and his people have passed from the land;
it is a book which makes him remembered
in the mouth of a speaker.
More excellent is a [papyrus] roll than a built house,
than a chapel in the west.
It is better than an established villa,
than a stela in a temple...

The use of writing in ancient Egypt was confined to the elite – the
royal family, courtiers, scribes, priests and some skilled workmen –
who formed less than five per cent of the population. To some extent
hieroglyphic texts must provide a biased view of ancient Egypt as
seen through their eyes, with only rare glimpses of what life was like

for the ordinary person, yet the situation was not greatly worse than that prevailing in Europe in Champollion's time when levels of literacy were never higher than fifty per cent and were usually a great deal lower – in some areas approaching the literacy levels of ancient Egypt. Despite any bias, the written texts provide some extraordinary insights into the culture of the ancient Egyptians, because of the variety of documents that has survived, whose contents could barely be imagined: deeds of sale, accounts, archives, tax documents, census lists, decrees, technical treatises, military despatches, lists of kings, funerary spells and rituals, letters to the living and to the dead, narrative tales – in fact almost every type of text found in modern societies, with the noticeable exception of dramatic works. The many notes that were jotted down on sherds of pottery and on small pieces of stone record trivial facts such as lists of building materials, who turned up for work on a particular day, or what was stored in a container. Probably used as teaching aids and works of reference, texts known as onomastica were lists of names by category, such as plants, animals, natural phenomena and even types of water: some Egyptian words are known only from such lists.

Egyptian society was focused on religion and hopes of an afterlife, subjects which particularly fascinated Champollion, and the relative abundance of some types of hieroglyphic texts reflects these values, as does the relatively good survival of some of the buildings, which was noticed by the savants who accompanied Napoleon to Egypt. In the account of his travels with the soldiers in the Nile Valley, the artist and author Vivant Denon noted that the ruins he saw were 'incessantly temples! No public edifice, not a house that has had strength to make a resistance to time, not a royal palace! Where, then, were the people? Where, then, were the sovereigns?' As temples and tombs were so important, they were constructed of stone to last as long as possible, but by contrast the homes of the Egyptian people, from the humblest peasant dwellings to the grand palaces of the pharaohs, were all built of mud brick, and very little now remains of these domestic structures. In their approach to death and the afterlife the Egyptians were strictly logical: a humble home or a palace could only be enjoyed for a lifetime, but a tomb was ⌐▭⌐⎯⎯◻ (*per n djet* – a house for eternity).

A single, unified religion never existed in ancient Egypt, because Egyptian religion developed from beliefs in many diverse gods with differing mythologies, but throughout Egypt the pharaoh was the intermediary between the gods and the people. At many temples, priests officiated on behalf of the pharaoh in rituals designed to maintain divine order and prevent the chaos that threatened to engulf it: a natural response to the situation in which the Egyptians found themselves. Life in the Nile Valley could be very pleasant when the forces of nature were peaceful, when the procession of the seasons was regular, and when there was no threat of war and invasion from neighbouring countries, but an unusually high or low annual flood from the Nile was enough to ruin crops and lead to drought and famine. The Egyptians were afraid of any change that might trigger some ruinous catastrophe, so their outlook was conservative, attempting to prevent change and to maintain stability. This essential harmony of the universe was personified by the goddess ⌐◻ (Maat), who also personified aspects of that harmony such as truth and justice, but many other gods and goddesses were believed to play a part in the prevention of chaos, which is why so many temples to different deities formed part of a single religion.

For the majority of the population, relationships with the gods were usually much more personal. Although in many ways still trying to maintain order and prevent chaos, individual worshippers looked to the gods for direct intervention in their lives. Local deities abounded, such as ⌐◻⎯◻◻◻ (Meretseger), the cobra goddess of the mountain overlooking the Valley of the Kings at Thebes, but there were also deities who were revered throughout Egypt, such as the goddess ◻◻ (Isis). On an even more basic level, household gods seldom had their own temples, but were worshipped in shrines within ordinary houses, where they were frequently called upon for help and protection. One of the most popular of these gods was 𓃀𓏤𓊃 (Bes), who was usually portrayed as a grotesquely ugly dwarf and whose fearsome appearance was believed to frighten away evil spirits. Thought to bring good luck and protect members of the household, Bes in particular was invoked for protection during childbirth – an especially dangerous event in ancient Egypt.

The modern distinctions between religious ritual, prayer, magic and science were not recognized by the ancient Egyptians, so that a doctor would often treat an illness with magic spells and rituals as well as medicines, each being seen as an essential part of the treatment. Amulets were commonly worn as a protection against specific evils, and Napoleon brought one back from Egypt which he always wore for good luck until his son was born. The following spell written down on papyrus was designed to be recited over a cornelian sealstone amulet (called here a 'knot'), engraved with images of a crocodile and a hand, to protect a child against fever:

Spell for a knot
for a child, a fledgling:
Are you hot in the nest?
Are you burning in the bush?
Your mother is not with you?
There is no sister there to fan you?
There is no nurse to offer protection?
Let there be brought to me a pellet of gold,
forty beads, a cornelian sealstone
with a crocodile and a hand on it
to fell, to drive off this Demon of Desire, to warm the limbs,
to fell these male and female enemies from the West.
You shall break out! This is a protection.
One shall say this spell over the pellet of gold,
the forty beads, a cornelian sealstone
with the crocodile and the hand.
To be strung on a strip of fine linen;

made into an amulet;
placed on the neck of the child.
Good.

Not only for the living, amulets were also bound up in the wrappings of mummies to protect the dead, and were frequently
inscribed with spells from the Book of the Dead, a modern name
given to what Champollion called 'the Funerary Ritual' and what
the Egyptians called the ⌒▯◯⊻⅄🕮 ▯◯ ᐱ◯ (Spells of Coming
Forth by Day). To ensure the survival of the dead in the afterlife
spells were recited by relatives or by priests, but in case they failed to
perform the rituals, steps were taken to give a more permanent
existence to the spells. For pharaohs from around 2300 B.C., the
texts of the spells were inscribed in hieroglyphs on the walls of the
chambers inside pyramids (spells now known as 'Pyramid Texts') to
ensure that they continued to be effective as long as the hieroglyphs
remained there. Around 2000 B.C. a change of ritual saw spells
being inscribed on the coffins of the deceased, rather than on the
walls of their tombs; the introduction of these 'Coffin Texts' coin-
cided with an increase in the number of people seeking survival
after death through mummification and magic spells. Originally
such treatment was the prerogative of the pharaoh, and courtiers
tried to be buried as close as possible to his tomb, hoping to share in
his resurrection, but now the elite had the spells written on their
own coffins in a direct attempt to secure their own survival after
death, without any association with the pharaoh. The spells tried to
cover every eventuality, from guarding against imprisonment in
the coffin to 'prevention of rotting and prevention of having to work
in the kingdom of the dead.'

About 500 years later spells now known as the Book of the Dead
began to supersede the Coffin Texts. Rather than being a book with
specific contents, the Book of the Dead comprised around 200
spells, many derived from the Pyramid Texts and Coffin Texts. Different copies of the Book of the Dead contained slightly varying sets

of spells, which were no longer written on the coffin, but on a roll of papyrus which was buried with the dead person in the tomb or inside the coffin, and many examples were first deciphered and studied by Champollion in the Drovetti collection at Turin. The need to make the spells for the protection of the dead durable, so that they would be effective for as long as possible, means that more copies of such spells have survived than any other type of Egyptian literature. Like the spells of the Pyramid Texts and the Coffin Texts, those of the Book of the Dead attempted to ensure the resurrection and survival of the person with whose body they were buried, but the way in which they were supposed to achieve this was complex. Instead of a body and soul, or a mind, body and spirit, the ancient Egyptians thought that a person consisted of five distinct elements: the physical body or corpse ⌒⋒ (*khat*); the 𓀓 (*ba*); the ⊔ (*ka*); the name of the person ⌒𓀓 (*ren*); and the person's shadow 𓏏𓏤 (*shut*).

The *ba*, sometimes roughly translated as 'personality,' was composed of all the non-physical elements that made a person unique. The *ka*, sometimes loosely translated as 'soul,' was regarded as the life force of an individual that lived on after death, but required sustenance. The food offerings to the dead person were not thought to be eaten and consumed by the *ka*, but the *ka* directly assimilated the life-preserving properties from the offerings. In order to survive after death, a person had to journey from the tomb and be reunited with his or her *ka*, but this being impossible for the physical body, it was undertaken by the *ba* instead. Once reunited, the *ba* and *ka* became the 𓅞 (*akh* – sometimes translated as 'the blessed dead'), the unchanging form in which the dead inhabited the underworld for eternity. The spells of the Book of the Dead were intended to ensure the proper formation of the *akh*, to protect it against all the dangers that threatened it in the afterlife, and to provide it with the most enjoyable type of life, free of work and care: its sinister modern title of 'Book of the Dead' is misleading, as the ancient Egyptian view of these collections of spells was closer to 'Book of Living Forever' or 'Book of Resurrection.'

Because of the belief that the worlds of the living and dead over-
lapped, letters were written to the dead on papyrus or on the bowls
containing their food offerings, but many other types of more mun-
dane letters were written by court officials, kings, priests and
artisans, some simply being short notes on potsherds. Ones dealing
with family business include those written to his family at Thebes
by Heqanakht, an elderly priest of the funeral cult of the dead Vizier
Ipi. He ends one letter by complaining about their poor treatment of
his new, second wife Iutenheb while he was away on business further down the Nile Valley and asks for her to be brought to him
(Hotepet is possibly a sister or aunt):

> I told you: 'Don't keep a friend of Hotepet away from her, nei-
> ther her hairdresser, nor her assistant.' Take great care of her!
> O may you prosper in everything accordingly! Yet, you have
> not loved her [Hotepet] in the past. You shall now cause
> Iutenheb to be brought to me. I swear by this man – I speak of
> Ipi – anyone who commits a misdeed against the sex of my
> new wife, he is against me and I against him. Look, this is my
> new wife; what should be done for a man's new wife is well
> known. Look, as for anyone who acts to her like I have acted –
> would one of you be patient when his wife was denounced to
> him? I'll be no more patient than you would!

All manner of documents could be found in temple archives,
including medical papyri, which provided advice on the diagnosis
and treatment of various medical conditions. One papyrus contains
the earliest known pregnancy test, which consisted of the woman
moistening grains of barley and emmer wheat with her urine every
day. If both types of grain germinated it indicated that she was
pregnant – if only the barley grew it indicated a male child, and if
only the emmer wheat grew it meant a female child. Modern experiments have shown that urine from a woman who is not pregnant
will prevent barley from growing, so there is some scientific support

for this ancient method. Other medical texts deal with the diagnosis and treatment of complaints such as broken bones, snakebites, bites from a range of other animals and diseases of the eye, and these give a vivid idea of the common hazards faced by the ancient Egyptians. Some texts provide more general moral advice, such as the 'Teaching of the Vizier Ptahhotep,' which is a collection of maxims such as:

> If you are a leader,
> be calm while you hear a petitioner's speech!
> Do not prevent him from purging his body
> of what he planned to tell you!
> A wronged man loves to pour his heart out
> more than achieving what he came for.
> About someone who prevents petitions, they say,
> 'So why does he thwart it?'
> Everything for which a man petitioned may not come about,
> but a good hearing is what soothes the heart.

Champollion believed that with the hieroglyphs he had deciphered the world's earliest known script and that writing had first developed in Egypt. Some decades after his death, the cuneiform script, originating in Mesopotamia, was finally deciphered, with some examples of cuneiform shown to be earlier than the earliest known hieroglyphs, making Mesopotamia the region where writing first developed. However, although there are very early types of cuneiform symbols that date to before the earliest hieroglyphs, the very latest finds in Egypt by archaeologists of examples of early hieroglyphs appear to show that Champollion may have been right after all – they are the earliest known instances of a *phonetic* script, dating from 3400 B.C.: the prototypes of the script that Champollion spent his life studying. At the very least, true writing developed simultaneously in both Egypt and Mesopotamia, but it is quite possible that it developed first in Egypt.

Even before he succeeded in deciphering the hieroglyphs, Champollion was aware that one of the most valuable prizes they could provide was a chronology for ancient Egypt because the texts covered such a huge range of time and often recorded historical events. Despite gaps and ambiguities, the Egyptian chronology still remains the most complete and most reliable of any country in the Mediterranean area before the advent of the Roman civilization. Because of the contacts that Egypt had with the Bronze Age cultures in Arabia, Anatolia and the Levant to the north-east, the Minoan and Mycenaean cultures of Greece and Crete, and the African cultures of Libya to the west and as far south as Sudan, Egypt has provided the backbone of ancient history in all these places: the decipherment of hieroglyphs not only unlocked Egypt's early history, but greatly enhanced the study of ancient history over an immense geographical area.

Today, Champollion is a national hero in France, but his contribution to world history is still only grudgingly recognized in some quarters, largely because the prejudices of his enemies and of Thomas Young's supporters had a disproportionate influence on scholars, especially in the English-speaking world, exaggerating Young's achievements and minimizing those of Champollion. A balanced view is seldom given, and the most important point is usually overlooked: irrespective of who first identified which particular hieroglyphic sign or who used what results of other researchers, Champollion's system of decipherment worked and Young's did not. In their efforts to promote Young as the true decipherer of hieroglyphs, his supporters did him immense harm as his real achievements in the decipherment of demotic are widely ignored.

In France many traces of Jean-François Champollion can be found, including statues, busts, pictures, commemorative plaques, street names, school names, memorials, a society and a museum. His portrait, painted by Léon Cogniet, is displayed in the Louvre Museum in Paris, not far from the Egyptian collections of which he was once curator. Most poignant are his statue by Frédéric Auguste Bartholdi in the courtyard of the College of France in Paris and his tomb in the Père Lachaise cemetery. Erected by his widow Rosine, the tomb is marked by a simple obelisk with the words 'Champollion le Jeune,' and railings surround a stone slab inscribed *'Ici repose Jean-François Champollion, né à Figeac dept. du Lot le 23 décembre 1790, décédé à Paris le 4 mars 1832'* (Here rests Jean-François Champollion, born at Figeac, Department of the Lot, on 23 December 1790, died at Paris on 4 March 1832). The house where he lived at 19 rue Mazarine can still be seen, and at number 28 in the same narrow street, where the crucial work on decipherment took place, Champollion is commemorated by a plaque, but the house at 4 rue Favart where he died has been demolished. Streets are named after him, with a rue Champollion in Paris and also in Grenoble, but there, in the town that was his home for so many years, massive redevelopment from the early twentieth century has swept away many of the buildings that he and Jacques-Joseph knew: only the lycée (now the lycée Stendhal) where Champollion was both pupil and teacher and the nearby entrance to the library where both brothers worked would still be readily recognized by them. A lycée Champollion (known to Grenoble people as 'Champo') was inaugurated in 1886.

Although he spent such a small part of his life there, Figeac is where Champollion's presence can now be felt most strongly. The house in the rue de la Boudousquerie (now the rue des frères Champollion) where Champollion and Jacques-Joseph were born is now a small but excellent museum, with a collection of Egyptian objects and a gallery devoted to the decipherment of hieroglyphs. His father's bookshop where he played as a child is currently the Sphinx café, but its upper storeys are little changed, and the square that witnessed both executions and celebrations during the Revolution has been renamed the Place Champollion (Champollion Square). Other buildings in the town also carry the name Champollion in his honour, including the modern lycée, and by the busy main road alongside the River Célé on the southern edge of the town an obelisk survives as a monument to Champollion, erected after his death by public subscription.

Apart from in Figeac, few memorials honour Jacques-Joseph Champollion-Figeac and his role in supporting Champollion during his life and after his death. After the revolution of February 1848 that caused the abdication of King Louis-Philippe, Jacques-Joseph lost his posts of Professor of Palaeography at the School of Charters (to which he had been appointed in 1830) and Keeper of Manuscripts at the Royal (National) Library, and he was then forced from his lodgings, having been accused of theft from the library. When in 1852 Napoleon III effected the transition from the Second Republic to the Second Empire, Jacques-Joseph regained favour and was appointed Keeper of the Library of the Palace of Fontainebleau. He kept this post until his death on 9 May 1867 at the age of eighty-nine (some thirty-five years after the death of his younger brother), and was buried in the cemetery at Fontainebleau. His son Ali had died in 1840, his wife Zoé in 1853 and his sons Jules and Paul in 1864 – only his son Aimé and daughter Zoé outlived him. Champollion's sister Marie died just one year after himself, in 1833, while Pétronille died fourteen years later and Thérèse at the end of 1851. The very modest family tomb in which Marie and Thérèse and their parents are buried is in the cemetery to the north of Figeac. The family house in the rue de la Boudousquerie was sold in 1854 and the bookshop the year after. Descendants of Jacques-Joseph still live in the Grenoble area, while Champollion's daughter Zoraïde married Amédée Chéronnet in 1845, and one of their children, René Chéronnet-Champollion, established an American branch of the family by marrying an American woman, Mary Corbin, and settling in New York.

One other memorial to Champollion initially appears a little bizarre – a crater on the moon is named after him. There is also a crater named after Joseph Fourier, the man who first introduced him to hieroglyphs, and one named after Champollion's main rival, Thomas Young. In a strange way this is a fitting tribute to these men who were so closely involved with the decipherment of hieroglyphs, because in ancient Egypt the moon was the province of the god Thoth who had several roles in Egyptian mythology, one of which was protector of the dead. The hieroglyph 𓅝 played a crucial role in the decipherment when Champollion recognized it as the symbol for the god Thoth in the name (Thothmes – born of the god Thoth), and in ancient Egyptian religion Thoth was identified with the moon and hailed as (Moon-Thoth). In some parts of Egypt the dead were thought to ride through the sky on the moon, protected by the god Thoth, but more importantly, throughout Egypt Thoth was the god of scribes and of knowledge and truth, and above all he was believed to have been the god who invented hieroglyphs – one of his many titles was (who gave words and script).

The ancient Egyptians believed that someone's name was an integral part of that person: to erase the name completely was to annihilate the person. Often vilified during his lifetime by his rivals and enemies who did their best to destroy his name and reputation, Champollion knew that his success would eventually come to be recognized. Just as the scribes of ancient Egypt were confident that their words would endure for ever, Champollion appreciated the wisdom of the ancient Egyptian maxim: 'It is good to speak to the Future; it shall listen.' His true memorial is that his name will be forever linked to the rediscovery of that civilization – the keys of ancient Egypt also opened the doors to Jean-François Champollion's own honoured place in history.

# Further Reading

You do not have to be able to read hieroglyphs to appreciate the treasures of ancient Egypt, but the more you know about the role of hieroglyphs, the greater is your enjoyment of them. Accessible introductions to the subject are *Understanding Hieroglyphs: A Complete Introductory Guide* by Hilary Wilson (1995), *ABC of Egyptian Hieroglyphs* by Jaromir Malek (1994) and *Egyptian Hieroglyphs* by W.V. Davies (1987). More detailed discussions of hieroglyphs can be found in *Hieroglyphs: The Writing of Ancient Egypt* by Maria Carmela Betrò (1996) and *Reading Egyptian Art: A Hieroglyphic Guide to Ancient Egyptian Painting and Sculpture* by Richard H. Wilkinson (1992). *Hieroglyphs and the Afterlife in Ancient Egypt* by Werner Forman and Stephen Quirke (1996) is a well-illustrated account of the historical development of hieroglyphs and *Papyrus* by Richard Parkinson and Stephen Quirke (1995) covers scribes and Egyptian scripts as well as the manufacture and use of papyrus.

If you wish to learn to read hieroglyphs, one of the best books is *How to Read Egyptian Hieroglyphs: A Step-by-step Guide to Teach Yourself* by Mark Collier and Bill Manley (1998). For a deeper study, *Egyptian Grammar: Being an Introduction to the Study of Hieroglyphs* by Alan Gardiner (revised 3rd edition, 1957) has yet to be superseded in its scope, although it is out of date on many points; James P. Allen, *Middle Egyptian: An Introduction to the Language and Culture of the Hieroglyphs* (2000) holds the best prospect so far of replacing it. No comprehensive hieroglyphic dictionaries exist in English, but the best available short dictionary is *A Concise Dictionary of Middle Egyptian* by Raymond O. Faulkner (1962). In German, a much more extensive dictionary is *Die Sprache der Pharaonen: Grosses Handwörterbuch Ägyptisch-Deutsch (2800-950 v. Chr)* by Rainer Hannig (1995).

For the latest word on hieroglyphs and the Rosetta Stone, *Cracking Codes: The Rosetta Stone and Decipherment* by Richard Parkinson (1999) is the best source, along with (in French) *La pierre de Rosette* by Robert Solé and Dominique Valbelle (1999), while more about pharaohs and their cartouches can be found in *Who were the Pharaohs? A History of their Names with a List of Cartouches* by Stephen Quirke (1990) and in *Chronicle of the Pharaohs: The Reign-by-reign Record of the Rulers and Dynasties of Ancient Egypt* by Peter Clayton (1994). Readable translations of hieroglyphic texts can be found in *Voices from Ancient Egypt: An Anthology of Middle Kingdom Writings* by R. B. Parkinson (1991) and in *The Tale of Sinuhe and Other Ancient Egyptian Poems 1940–1640 B.C.* also by R. B. Parkinson (1997). The translations in the series by Miriam Lichtheim, *Ancient Egyptian Literature: A Book of Readings, Volume I: The Old and Middle Kingdoms* (1973); *Ancient Egyptian Literature: A Book of Readings, Volume II: The New Kingdom* (1976); and *Ancient Egyptian Literature: A Book of Readings, Volume III: The Late Period* (1980), provide an overview of the variety of ancient Egyptian literature. *The Ancient Egyptian Book of the Dead* translated by Raymond O. Faulkner and edited by Carol Andrews (revised edition 1985) provides an illustrated version of this text.

Several books have been written on Napoleon's campaign in Egypt, of which *Bonaparte in Egypt* by J. Christopher Herold (1962) is one of the best, as well as (in French) *L'expédition d'Égypte 1798–1801* by Henry Laurens (1997) and *L'Égypte, une aventure savante 1798–1801* by Yves Laissus (1998). Vivant Denon's *Travels in Upper and Lower Egypt during the Campaigns of General Bonaparte* (translated from the French in 1802) was reprinted in 1986. There are no accounts or biographies of Champollion in English: almost all the source material is in French, although one biography was first written in German – *Champollion, sein Leben und sein Werk* by H. Hartleben (1906, in two volumes), abridged and translated into French by Denise Meunier as *Jean-François Champollion: Sa vie et son oeuvre 1790–1832* (1983). A more recent biography in French is *Champollion: Une vie de lumières* by Jean Lacouture (1988). For Thomas Young, the most recent and accessible source is *Thomas Young Natural Philosopher 1773–1829* by Alexander Wood and Frank Oldham (1954), and accounts of some of the failed attempts to decipher hieroglyphs by Champollion's rivals and earlier researchers are given in *The Myth of Egypt and its Hieroglyphs in European Tradition* by Erik Iversen (1961, reprinted 1993).

There are numerous well-illustrated books to choose from on most aspects of ancient Egypt itself: the *British Museum Dictionary of Ancient Egypt* by Ian Shaw and Paul Nicholson (1994) is a useful reference book, while *The Mummy in Ancient Egypt: Equipping the Dead for Eternity* by Salima Ikram and Aidan Dodson (1998) has the subject wrapped up. The pyramids are explained in *The Complete Pyramids* by Mark Lehner (1997) and *The Pyramids* by Alberto Siliotti (1997). The impressive sites of Thebes (at one time the capital of ancient Egypt) are described in an accessible style in *Thebes in Egypt: A Guide to the Tombs and Temples of Ancient Luxor* by Nigel and Helen Strudwick (1999), while the Valley of the Kings is the subject of the *Guide to the Valley of the Kings* by Alberto Siliotti (1996) and *The Complete Valley of the Kings* by Nicholas Reeves and Richard H. Wilkinson (1996). Sites throughout Egypt and Nubia are described and mapped in *Atlas of Ancient Egypt* by John Baines and Jaromír Málek (1984). Finally, an idea of the richness of Egyptian civilization can be seen in the pages of *Egypt: The World of the Pharaohs* edited by Regine Schulz and Matthias Seidel (1998).

# Index

friendships 67, 92, 98, 139,
144, 163–4, 176
ill-health 156, 160, 205;
bronchial problems 79, 141,
283–4; exhaustion 68, 155,
253, 269; eyesight problems
68; final illness 286–7, 289;
gout 236, 257, 261, 279,
284, 286, 289; headaches 79;
physical collapse 181, 182,
186, 187; rheumatism 217;
suffers stroke 286, 289;
weight loss 79, 141
in Italy 204, 206–7, 209, 210,
211–27, 230–6
letters to brother: from Egypt
243–4, 246–7, 249, 252,
254, 256, 258, 260–2, 265,
267, 269, 271–2; on change
of career 134; on poor health
205, 279, 285; on progress
with hieroglyphs 85, 145–6;
from Italy 211–12, 216, 218,
220–1, 223–4, 226, 229;
from Paris 73, 74–5, 77, 79;
from school 55, 67–70
lion motif of 45, 174
London visit 209
love affairs 105, 230–1
marriage 147
memorials to 306–8
parents and family 2, 42–4, 56,
137, 141, 308
poverty 55–6, 79–81, 100,
105, 109, 198, 204–5
rejects change of surname 49
relations with brother 55–6, 70,
71, 80–1, 127
relations with wife 147–8
religious beliefs 199, 223
student days 72–89
tomb 307
personality and intellect:

aversion to mathematics 46,
50–1
drawing skills 3, 4, 45, 51
flair for making enemies 67, 99,
134–5, 190, 224, 228–9, 231
focus and tenacity 75
intellectual curiosity 2–3
intellectual gifts 3, 4–5, 44, 45,
46, 75
mood swings 4, 45, 47, 49, 54
quick temper 45
self-confidence 87, 203
strength of character 45
visual memory 3
politics:
attitude to monarchy 111, 204
charged with treason 159–60
exiled to Figeac 135–42, 147
involvement in Grenoble
uprising (1821)
passion for freedom 56
political enemies 100–1, 134,
141, 144, 156, 160, 164,
179, 207, 209, 238
political satires 126
republicanism 107–8, 128
seeks royal patronage 204,
205–7
support for Napoleonic regime
126–8, 134
views on Napoleon 5, 73, 108
professional life:
assessment of rivals 87–9, 95–
6, 191–2
collaborates with Biot on
Egyptian calendar 282
contacts with Young 113, 122,
125, 190, 203, 240–1
curator in Louvre 231, 235,
236, 238–9
diplomatic adviser 145
educational activities 140–1,
144–5, 146, 155

number of signs 208
numerical system 132
phonetic 83–4, 151–3, 174–5,
183–7, 191, 208
phonetic complements 183–4
pictograms 131, 183
plurals 131
previous attempts to decipher 57,
59–64, 87, 96, 223
Rosetta Stone text 35, 38, 62–3,
89, 95, 130, 151–2, 171
symbolism of 57–8, 59–60, 69,
86, 96
translations of 296
uniliterals, biliterals and triliterals
184–5, 294
word endings 105
Young's theory of 131, 150, 153–
4
Hincks, Samuel 294
Horapollo 57, 69
Horemheb, Pharaoh 261
Humboldt, Alexander von 160
Huyot, Jean-Nicolas 180, 258

Ibrahim Bey 19
*Iliad* (Homer) 265
Institute of Egypt, Cairo, *see*
Egyptian Institute of Arts and
Sciences
Institute of France, Paris 1, 17, 74,
181, 280
Academies 163
acquires copy of Rosetta Stone 38
Young's membership 117, 276
*see also under individual academies*
*Introduction to Medical Literature*
(Young) 118
Isabella, Queen of Naples 220
Italy 145
Champollion visits 211–27, 230–
6
Egyptian collections 204, 206

Egyptology in 231
hostilities with France 116, 138
Renaissance 57–8, 59–60
Iutenheb 304

Jacobins 100–1
Jacqou the Sorcerer 43
Jollois, Prosper 33, 34, 78, 165
Jomard, Edme-François 10, 39, 78,
88
Champollion's assessment of 176,
252–3, 260, 274
collaboration with Champollion
238
correspondence with Young 129–
30, 154
and Dendera zodiac 166, 167,
274
edits *Description de l'Égypte* 78,
130, 163, 176, 198
hostility towards Champollion
78, 146, 164, 198–9, 267,
294
lobbies for purchase of
Passalacqua collection 228–9
presentation to Academy 182
relations with Jacques-Joseph
78
tries to help Champollion avoid
conscription 139
work on hieroglyphic numerals
132
Josephine Bonaparte (Marie-
Joseph-Rose de Beauharnais)
decoration of country house 86
divorce 93
during Revolution 48, 49
and Egyptian expedition 7–8
as Empress 53
ill-health 8
marriage 8, 48
patronage of Denon 29
relations with Napoleon 8, 93